Aeronau
Engineer
Data Bo

Clifford Matthews BSc, CEng, MBA, FIMechE

BUTTERWORTH
HEINEMANN

OXFORD AUCKLAND BOSTON JOHANNESBURG
MELBOURNE NEW DELHI

Butterworth-Heineman
Linacre House, Jordan Hill, Oxford OX2 8DP
225 Wildwood Avenue, Woburn, MA 01801-2041
A division of Reed Educational and Professional Publishing Ltd

A member of the Reed Elsevier plc group

First published 2002

British Library Cataloguing in Publication Data
Matthews, Clifford
Aeronautical engineer's data book
1. Aerospace engineering–Handbooks, manuals, etc.
I. Title
629.1'3

Library of Congress Cataloguing in Publication Data
Matthews, Clifford.
Aeronautical engineer's data book / Clifford Matthews.
p. cm.
Includes index.
ISBN 0 7506 5125 3
1. Aeronautics–Handbooks, Manuals, etc. I. Title.
TL570.M34 2001
629.13'002'12–dc21 2001037429

ISBN 0 7506 5125 3

Composition by Scribe Design, Gillingham, Kent, UK
Transferred to digital print 2008
Printed and bound in Great Britain by
CPI Antony Rowe, Eastbourne

Contents

Preface

The objective of this *Aeronautical Engineer's Data book* is to provide a concise and useful source of up-to-date information for the student or practising aeronautical engineer. Despite the proliferation of specialized information sources, there is still a need for basic data on established engineering rules, conversions, modern aircraft and engines to be available in an easily assimilated format.

An aeronautical engineer cannot afford to ignore the importance of engineering data and rules. Basic theoretical principles underlie the design of all the hardware of aeronautics. The practical processes of fluid mechanics, aircraft design, material choice, and basic engineering design form the foundation of the subject. Technical standards, directives and regulations are also important – they represent accumulated knowledge and form invaluable guidelines for the industry.

The purpose of the book is to provide a basic set of technical data that you will find useful. It is divided into 13 sections, each containing specific 'discipline' information. Units and conversions are covered in Section 2; a mixture of metric and imperial units are still in use in the aeronautical industry. Information on FAA regulations is summarized in Section 1 – these develop rapidly and affect us all. The book contains cross-references to other standards systems and data sources. You will find these essential if you need to find more detailed information on a particular subject. There is always a limit to the amount of information that you can carry with you – the secret is knowing where to look for the rest.

More and more engineering information is now available in electronic form and many engineering students now use the Internet as their first source of reference information for technical information. This new *Aeronautical Engineer's Data Book* contains details of a wide range of engineering-related websites, including general 'gateway' sites such as the Edinburgh Engineering Virtual Library (EEVL) which contains links to tens of thousands of others containing technical information, product/company data and aeronautical-related technical journals and newsgroups.

You will find various pages in the book contain 'quick guidelines' and 'rules of thumb'. Don't expect these all to have robust theoretical backing – they are included simply because I have found that they *work*. I have tried to make this book a practical source of aeronautics-related technical information that you can use in the day-to-day activities of an aeronautical career.

Finally, it is important that the content of this data book continues to reflect the information that is needed and used by student and experienced engineers. If you have any suggestions for future content (or indeed observations or comment on the existing content) please submit them to me at the following e-mail address: aerodatabook@aol.com

Clifford Matthews

Acknowledgements

Special thanks are due to Stephanie Evans, Sarah Pask and John King for their excellent work in typing and proof reading this book.

Disclaimer

This book is intended to assist engineers and designers in understanding and fulfilling their obligations and responsibilities. All interpretation contained in this publication – concerning technical, regulatory and design information and data, unless specifically otherwise identified, carries no authority. The information given here is not intended to be used for the design, manufacture, repair, inspection or certification of aircraft systems and equipment, whether or not that equipment is subject to design codes and statutory requirements. Engineers and designers dealing with aircraft design and manufacture should not use the information in this book to demonstrate compliance with any code, standard or regulatory requirement. While great care has been taken in the preparation of this publication, neither the author nor the publishers do warrant, guarantee, or make any representation regarding the use of this publication in terms of correctness, accuracy, reliability, currentness, comprehensiveness, or otherwise. Neither the publisher, author, nor anyone, nor anybody who has been involved in the creation, production, or delivery of this product shall be liable for any direct, indirect, consequential, or incidental damages arising from its use.

Section 1

Important regulations and directives

A fundamental body of information is contained in the US Federal Aviation Regulations (FAR). A general index is shown below:

Federal Aviation Regulations

Chapters I and III

Subchapter A – definitions and abbreviations

Part 1: Definitions and abbreviations

Subchapter B – procedural rules

Part 11: General rule-making procedures
Part 13: Investigative and enforcement procedures
Part 14: Rules implementing the Equal Access to Justice Act of 1980
Part 15: Administrative claims under Federal Tort Claims Act
Part 16: Rules of practice for federally-assisted airport enforcement proceedings
Part 17: Procedures for protests and contracts disputes

Subchapter C – aircraft

Part 21: Certification procedures for products and parts
Part 23: Airworthiness standards: normal, utility, acrobatic, and commuter category airplanes
Part 25: Airworthiness standards: transport category airplanes

Part 27: Airworthiness standards: normal category rotorcraft
Part 29: Airworthiness standards: transport category rotorcraft
Part 31: Airworthiness standards: manned free balloons
Part 33: Airworthiness standards: aircraft engines
Part 34: Fuel venting and exhaust emission requirements for turbine engine powered airplanes
Part 35: Airworthiness standards: propellers
Part 36: Noise standards: aircraft type and airworthiness certification
Part 39: Airworthiness directives
Part 43: Maintenance, preventive maintenance, rebuilding, and alteration
Part 45: Identification and registration marking
Part 47: Aircraft registration
Part 49: Recording of aircraft titles and security documents

Subchapter D – airmen

Part 61: Certification: pilots and flight instructors
Part 63: Certification: flight crewmembers other than pilots
Part 65: Certification: airmen other than flight crewmembers
Part 67: Medical standards and certification

Subchapter E – airspace

Part 71: Designation of class a, class b, class c, class d, and class e airspace areas; airways; routes; and reporting points
Part 73: Special use airspace
Part 77: Objects affecting navigable airspace

Subchapter F – air traffic and

general operation rules

Part 91: General operating and flight rules
Part 93: Special air traffic rules and airport traffic patterns
Part 95: IFR altitudes
Part 97: Standard instrument approach procedures
Part 99: Security control of air traffic
Part 101: Moored balloons, kites, unmanned rockets and unmanned free balloons
Part 103: Ultralight vehicles
Part 105: Parachute jumping
Part 107: Airport security
Part 108: Airplane operator security
Part 109: Indirect air carrier security

Subchapter G – air carriers and operators for compensation or hire: certification and operations

Part 119: Certification: air carriers and commercial operators
Part 121: Operating requirements: domestic, flag, and supplemental operations
Part 125: Certification and operations: airplanes having a seating capacity of 20 or more passengers or a maximum payload capacity of 6000 pounds or more
Part 129: Operations: foreign air carriers and foreign operators of US – registered aircraft engaged in common carriage
Part 133: Rotorcraft external-load operations
Part 135: Operating requirements: commuter and on-demand operations
Part 137: Agricultural aircraft operations
Part 139: Certification and operations: land airports serving certain air carriers

Subchapter H – schools and other certificated agencies

Part 141: Pilot schools
Part 142: Training centers
Part 145: Repair stations
Part 147: Aviation maintenance technician schools

Subchapter I – airports

Part 150: Airport noise compatibility planning
Part 151: Federal aid to airports
Part 152: Airport aid program
Part 155: Release of airport property from surplus property disposal restrictions
Part 156: State block grant pilot program
Part 157: Notice of construction, alteration, activation, and deactivation of airports
Part 158: Passenger Facility Charges (PFCs)
Part 161: Notice and approval of airport noise and access restrictions
Part 169: Expenditure of federal funds for nonmilitary airports or air navigation facilities thereon

Subchapter J – navigational facilities

Part 170: Establishment and discontinuance criteria for air traffic control services and navigational facilities
Part 171: Non-federal navigation facilities

Subchapter K – administrative regulations

Part 183: Representatives of the administrator
Part 185: Testimony by employees and production of records in legal proceedings, and service of legal process and pleadings
Part 187: Fees
Part 189: Use of federal aviation administration communications system

Part 191: Withholding security information from disclosure under the Air Transportation Security Act of 1974

Subchapter N – war risk insurance

Part 198: Aviation insurance

Chapter III – parts 400 to 440

Subchapter A – general

Part 400: Basis and scope
Part 401: Organization and definitions

Subchapter B – procedure

Part 404: Regulations and licensing requirements
Part 405: Investigations and enforcement
Part 406: Administrative review

Subchapter C – licensing

Part 413: Applications
Part 415: Launch licenses
Part 417: License to operate a launch site
Part 440: Financial responsibility

Requests for information or policy concerning a particular Federal Aviation Regulation should be sent to the office of primary interest (OPI). Details can be obtained from FAA's consumer hotline, in the USA toll free, at 1-800-322-7873.

Requests for interpretations of a Federal Aviation Regulation can be obtained from:

Federal Aviation Administration
800 Independence Ave SW
Washington, DC 20591
USA

Section 2

Fundamental dimensions and units

2.1 The Greek alphabet

The Greek alphabet is used extensively in Europe and the United States to denote engineering quantities (see Table 2.1). Each letter can have various meanings, depending on the context in which it is used.

Table 2.1 The Greek alphabet

Name	*Symbol*	
	Capital	*Lower case*
alpha	A	α
beta	B	β
gamma	Γ	γ
delta	Δ	δ
epsilon	E	ϵ
zeta	Z	ζ
eta	H	η
theta	Θ	θ
iota	I	ι
kappa	K	κ
lambda	Λ	λ
mu	M	μ
nu	N	ν
xi	Ξ	ξ
omicron	O	ο
pi	Π	π
rho	P	ρ
sigma	Σ	σ
tau	T	τ
upsilon	Y	υ
phi	Φ	ϕ
chi	X	χ
psi	Ψ	ψ
omega	Ω	ω

2.2 Units systems

The most commonly used system of units in the aeronautics industry in the United States is the United States Customary System (USCS). The 'MKS system' is a metric system still used in some European countries but is gradually being superseded by the expanded Système International (SI) system.

2.2.1 The USCS system

Countries outside the USA often refer to this as the 'inch-pound' system. The base units are:

Length: foot (ft) = 12 inches (in)
Force: pound force or thrust (lbf)
Time: second (s)
Temperature: degrees Fahrenheit (°F)

2.2.2 The SI system

The strength of the SI system is its *coherence*. There are four mechanical and two electrical base units from which all other quantities are derived. The mechanical ones are:

Length: metre (m)
Mass: kilogram (kg)
Time: second (s)
Temperature: Kelvin (K) or, more commonly, degrees Celsius or Centigrade (°C)

Other units are derived from these: e.g. the newton (N) is defined as $N = kg\ m/s^2$. Formal SI conversion factors are listed in ASTM Standard E380.

2.2.3 SI prefixes

As a rule, prefixes are generally applied to the basic SI unit, except for weight, where the prefix is used with the unit gram (g), not the basic SI unit kilogram (kg). Prefixes are not used for units of angular measurement (degrees, radians), time (seconds) or temperature (°C or K).

Prefixes are generally chosen in such a way that the numerical value of a unit lies between 0.1 and 1000 (see Table 2.2). For example:

28 kN rather than 2.8×10^4 N
1.25 mm rather than 0.00125 m
9.3 kPa rather than 9300 Pa

Table 2.2 SI unit prefixes

Multiplication factor	*Prefix*	*Symbol*
1 000 000 000 000 000 000 000 000 = 10^{24}	yotta	Y
1 000 000 000 000 000 000 000 = 10^{21}	zetta	Z
1 000 000 000 000 000 000 = 10^{18}	exa	E
1 000 000 000 000 000 = 10^{15}	peta	P
1 000 000 000 000 = 10^{12}	tera	T
1 000 000 000 = 10^{9}	giga	G
1 000 000 = 10^{6}	mega	M
1 000 = 10^{3}	kilo	k
100 = 10^{2}	hicto	h
10 = 10^{1}	deka	da
0.1 = 10^{-1}	deci	d
0.01 = 10^{-2}	centi	c
0.001 = 10^{-3}	milli	m
0.000 001 = 10^{-6}	micro	μ
0.000 000 001 = 10^{-9}	nano	n
0.000 000 000 001 = 10^{-12}	pico	p
0.000 000 000 000 001 = 10^{-15}	femto	f
0.000 000 000 000 000 001 = 10^{-18}	atto	a
0.000 000 000 000 000 000 001 = 10^{-21}	zepto	z
0.000 000 000 000 000 000 000 001 = 10^{-24}	yocto	y

2.3 Conversions

Units often need to be converted. The least confusing way to do this is by expressing *equality*:

For example, to convert 600 lb thrust to kilograms (kg)
Using 1 kg = 2.205 lb

Add denominators as

$$\frac{1\ \text{kg}}{x} = \frac{2.205\ \text{lb kg}}{600\ \text{lb}}$$

Solve for x

$$x = \frac{600 \times 1}{2.205} = 272.1\ \text{kg}$$

Hence 600 lb = 272.1 kg

Setting out calculations in this way can help avoid confusion, particularly when they involve large numbers and/or several sequential stages of conversion.

2.3.1 Force or thrust

The USCS unit of force or thrust is the *pound force* (*lbf*). Note that a pound is also ambiguously used as a unit of mass (see Table 2.3).

Table 2.3 Force (F) or thrust

Unit	*lbf*	*gf*	*kgf*	*N*
1 pound thrust (lbf)	1	453.6	0.4536	4.448
1 gram force (gf)	2.205×10^{-3}	1	0.001	9.807×10^{-3}
1 kilogram-force (kgf)	2.205	1000	1	9.807
1 newton (N)	0.2248	102.0	0.1020	1

Note: Strictly, all the units in the table except the newton (N) represent weight equivalents of mass and so depend on the 'standard' acceleration due to gravity (g). The true SI unit of force is the newton (N) which is equivalent to 1 kgm/s^2.

2.3.2 Weight

The true weight of a body is a measure of the gravitational attraction of the earth on it. Since this attraction is a force, the weight of a body is correctly expressed in pounds force (lbf).

Mass is measured in pounds mass (lbm) or simply (lb)
Force (lbf) = mass (lbm) $\times$ g (ft/s^2)
Or, in SI units: force (N) = mass (kg) $\times$ g (m/s^2)
1 ton (US) = 2000 lb = 907.2 kg
1 tonne (metric) = 1000 kg = 2205 lb

2.3.3 Density

Density is defined as mass per unit volume. Table 2.4 shows the conversions between units.

Table 2.4 Density (ρ)

Unit	*lb/in³*	*lb/ft³*	*kg/m³*	*g/cm³*
1 lb per in³	1	1728	2.768×10^4	27.68
1 lb per ft³	5.787×10^{-4}	1	16.02	1.602×10^{-2}
1 kg per m³	3.613×10^{-5}	6.243×10^{-2}	1	0.001
1 g per cm³	3613×10^{-2}	62.43	1000	1

2.3.4 Pressure

The base USCS unit is the lbf/in² (or 'psi').

1 Pa = 1 N/m²
1 Pa = 1.45038×10^{-4} lbf/in²

In practice, pressures in SI units are measured in MPa, bar, atmospheres, torr, or the height of a liquid column, depending on the application. See Figures 2.1, 2.2 and Table 2.5.

So for liquid columns:

1 in H_2O = 25.4 mm H_2O = 249.089 Pa
1 in Hg = 13.59 in H_2O = 3385.12 Pa = 33.85 mbar.
1 mm Hg = 13.59 mm H_2O = 133.3224 Pa = 1.333224 mbar.
1 mm H_2O = 9.80665 Pa
1 torr = 133.3224 Pa

For conversion of liquid column pressures: 1 in = 25.4 mm.

2.3.5 Temperature

The basic unit of temperature is degrees Fahrenheit (°F). The SI unit is kelvin (K). The most commonly used unit is degrees Celsius (°C).

Absolute zero is defined as 0 K or –273.15°C, the point at which a perfect gas has zero volume. See Figures 2.3 and 2.4.

$°C = {}^5/_9\,(°F - 32)$
$°F = {}^9/_5\,(°C + 32)$

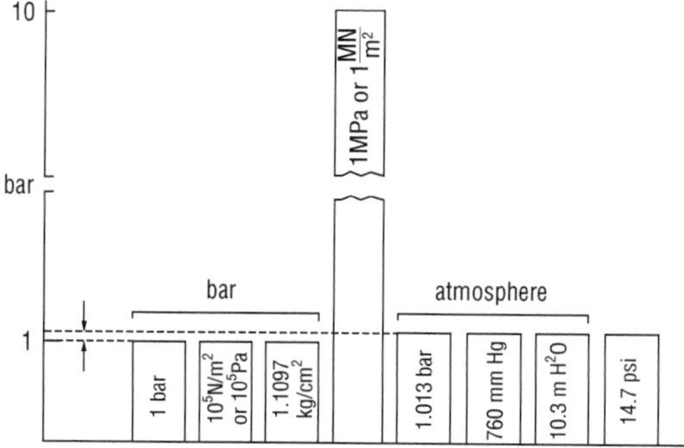

10
bar
1
1MPa or 1 MN/m²
bar
atmosphere
1 bar
10⁵N/m² or 10⁵Pa
1.1097 kg/cm²
1.013 bar
760 mm Hg
10.3 m H²O
14.7 psi
Rules of thumb: An apple 'weighs' about 1.5 newtons
A meganewton is equivalent to about 100 tonnes
An average car weighs about 15 kN

Fig. 2.1 Pressure relationships

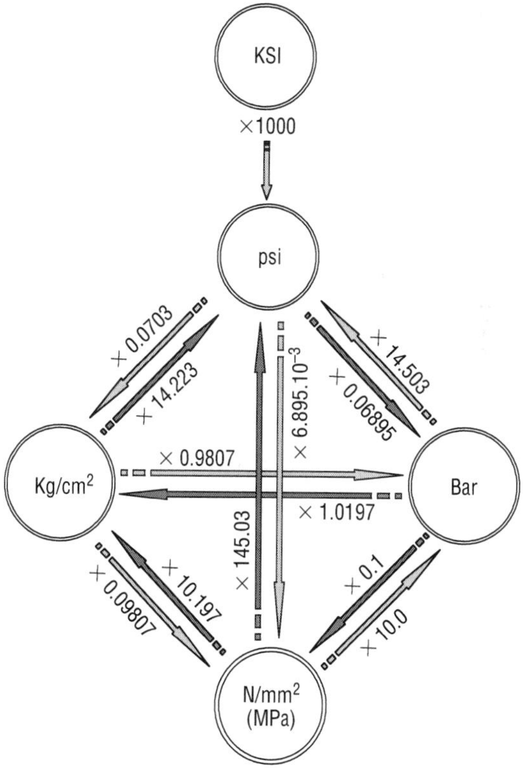

KSI
×1000
psi
× 0.0703
× 14.223
× 6.895.10⁻³
× 14.503
× 0.06895
Kg/cm²
× 0.9807
× 1.0197
Bar
× 145.03
× 0.09807
× 10.197
× 0.1
× 10.0
N/mm² (MPa)

Fig. 2.2 Pressure conversions

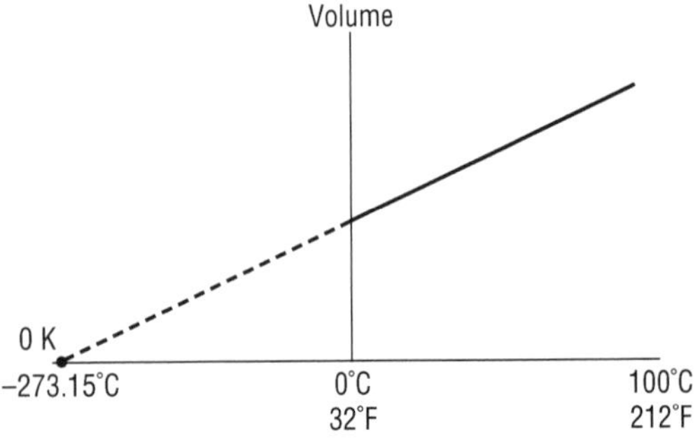

Fig. 2.3 Temperature

2.3.6 Heat and work

The basic unit for heat 'energy' is the British thermal unit (BTU).

Specific heat 'energy' is measured in BTU/lb (in SI it is joules per kilogram (J/kg)).

1 J/kg = 0.429923×10^{-3} BTU/lb

Table 2.6 shows common conversions.

Specific heat is measured in BTU/lb °F (or in SI, joules per kilogram kelvin (J/kg K)).

1 BTU/lb °F = 4186.798 J/kg K
1 J/kg K = 0.238846 (10^{-3} BTU/lb °F
1 kcal/kg K = 4186.8 J/kg K

Heat flowrate is also defined as power, with the unit of BTU/h (or in SI, in watts (W)).

1 BTU/h = 0.07 cal/s = 0.293 W
1 W = 3.41214 BTU/h = 0.238846 cal/s

2.3.7 Power

BTU/h or horsepower (hp) are normally used or, in SI, kilowatts (kW). See Table 2.7.

2.3.8 Flow

The basic unit of volume flowrate is US gallon/min (in SI it is litres/s).

1 US gallon = 4 quarts = 128 US fluid ounces = 231 in^3

1 US gallon = 0.8 British imperial gallons = 3.78833 litres
1 US gallon/minute = 6.31401×10^{-5} m^3/s = 0.2273 m^3/h
1 m^3/s = 1000 litres/s
1 litre/s = 2.12 ft^3/min

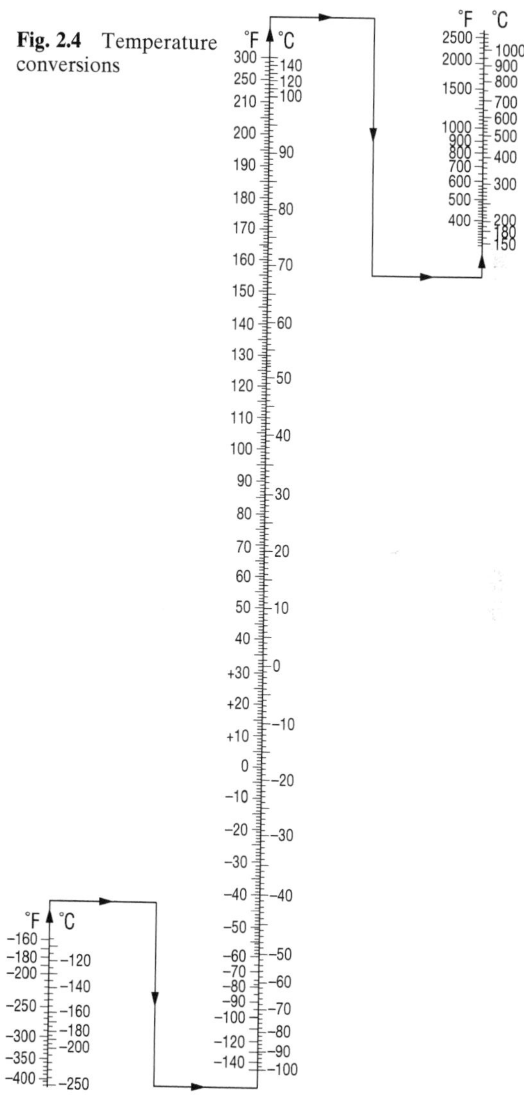

Fig. 2.4 Temperature conversions

Table 2.5 Pressure (p)

Unit	*lb/in² (psi)*	*lb/ft²*	*atm*	*in H_20*	*cmHg*	*N/m²(Pa)*
1 lb per in² (psi)	1	144	6.805×10^{-2}	27.68	5.171	6.895×10^{3}
1 lb per ft²	6.944×10^{-3}	1	4.725×10^{-4}	0.1922	3.591×10^{-2}	47.88
1 atmosphere (atm)	14.70	2116	1	406.8	76	1.013×10^{5}
1 in of water at 39.2°F (4°C)	3.613×10^{-2}	5.02	2.458×10^{-3}	1	0.1868	249.1
1 cm of mercury at 32°F (0°C)	0.1934	27.85	1.316×10^{-2}	5.353	1	1333
1 N per m² (Pa)	1.450×10^{-4}	2.089×10^{-2}	9.869×10^{-6}	4.015×10^{-3}	7.501×10^{-4}	1

Table 2.6 Heat

	BTU	*ft-lb*	*hp-h*	*cal*	*J*	*kW-h*
1 British thermal unit (BTU)	1	777.9	3.929×10^{-4}	252	1055	2.93×10^{-4}
1 foot-pound (ft-lb)	1.285×10^{-3}	1	5.051×10^{-7}	0.3239	1.356	3.766×10^{-7}
1 horsepower-hour (hp-h)	2545	1.98×10^{6}	1	6.414×10^{5}	2.685×10^{6}	0.7457
1 calorie (cal)	3.968×10^{-3}	3.087	1.559×10^{-6}	1	4.187	1.163×10^{-6}
1 joule (J)	9.481×10^{-4}	0.7376	3.725×10^{-7}	0.2389	1	2.778×10^{-7}
1 kilowatt hour (kW-h)	3413	2.655×10^{6}	1.341	8.601×10^{5}	3.6×10^{6}	1

Table 2.7 Power (P)

	BTU/h	*BTU/s*	*ft-lb/s*	*hp*	*cal/s*	*kW*	*W*
1 BTU/h	1	2.778×10^{-4}	0.2161	3.929×10^{-4}	7.000×10^{-2}	2.930×10^{-4}	0.2930
1 BTU/s	3600	1	777.9	1.414	252.0	1.055	1.055×10^{-3}
1ft-lb/s	4.62	1.286×10^{-3}	1	1.818×10^{-3}	0.3239	1.356×10^{-3}	1.356
1 hp	2545	0.7069	550	1	178.2	0.7457	745.7
1 cal/s	14.29	0.3950	3.087	5.613×10^{-3}	1	4.186×10^{-3}	4.186
1 kW	3413	0.9481	737.6	1.341	238.9	1	1000
1 W	3.413	9.481×10^{-4}	0.7376	1.341×10^{-3}	0.2389	0.001	1

Table 2.8 Velocity (v)

Item	*ft/s*	*km/h*	*m/s*	*mile/h*	*cm/s*	*knot*
1 ft per s	1	1.097	0.3048	0.6818	30.48	0.592
1 km per h	0.9113	1	0.2778	0.6214	27.78	0.5396
1 m per s	3.281	3.600	1	2.237	100	1.942
1 mile per h	1.467	1.609	0.4470	1	44.70	0.868
1 cm per s	3.281×10^{-2}	3.600×10^{-2}	0.0100	2.237×10^{-2}	1	0.0194
1 knot	1.689	1.853	0.5148	1.152	51.48	1

2.3.9 Torque

The basic unit of torque is the foot pound (ft.lbf) (in SI it is the newton metre (N m)). You may also see this referred to as 'moment of force' (see Figure 2.5)

1 ft.lbf= 1.357 N m
1 kgf.m = 9.81 N m

2.3.10 Stress

Stress is measured in lb/in^2 – the same unit used for pressure although it is a different physical quantity. In SI the basic unit is the pascal (Pa). 1 Pa is an impractically by small unit so MPa is normally used (see Figure 2.6).

1 lb/in^2 = 6895 Pa
1 MPa = 1 MN/m^2 = 1 N/mm^2
1 kgf/mm^2 = 9.80665 MPa

2.3.11 Linear velocity (speed)

The basic unit of linear velocity (speed) is feet per second (in SI it is m/s). In aeronautics, the most common non-SI unit is the knot, which is equivalent to 1 nautical mile (1853.2 m) per hour. See Table 2.8.

2.3.12 Acceleration

The basic unit of acceleration is feet per second squared (ft/s^2). In SI it is m/s^2.

1 ft/s^2 = 0.3048 m/s^2

1 m/s^2 = 3.28084 ft/s^2

Standard gravity (g) is normally taken as 32.1740 ft/s^2 (9.80665 m/s^2).

2.3.13 Angular velocity

The basic unit is radians per second (rad/s).

1 rad/s = 0.159155 rev/s = 57.2958 degree/s

The radian is also the SI unit used for plane angles.

A complete circle is 2π radians (see Figure 2.7)
A quarter-circle (90°) is $\pi/2$ or 1.57 radians
1 degree = $\pi/180$ radians

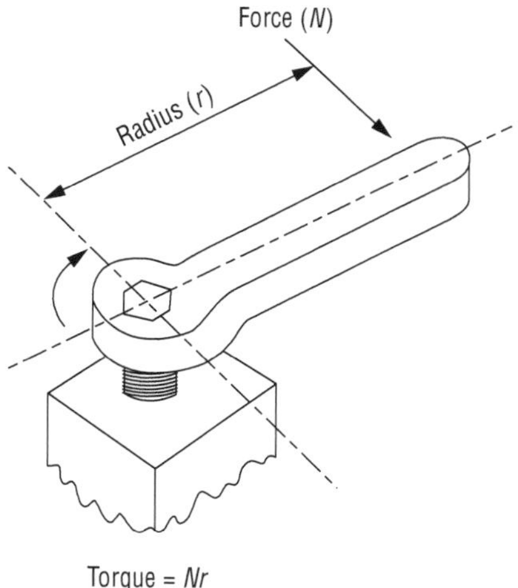

Fig. 2.5 Torque

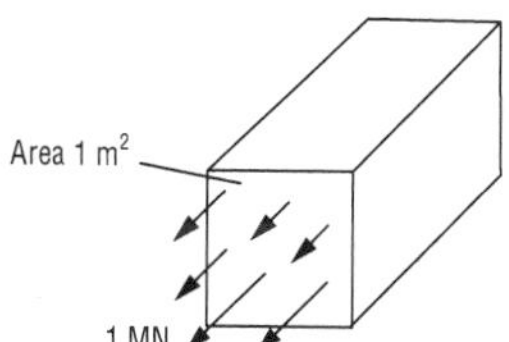

Fig. 2.6 Stress

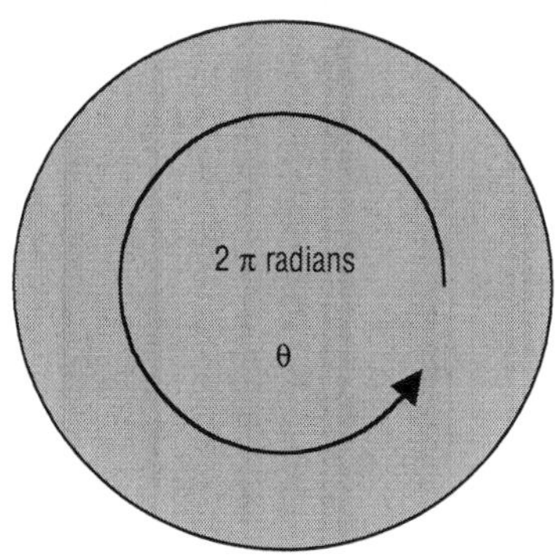

Fig. 2.7 Angular measure

Table 2.9 Area (A)

Unit	*sq.in*	*sq.ft*	*sq.yd*	*sq.mile*	cm^2	dm^2	m^2	*a*	*ha*	km^2
1 square inch	1	-	–	–	6.452	0.06452	–	-	–	-
1 square foot	144	1	0.1111	-	929	9.29	0.0929	–	-	–
1 square yard	1296	9	1	–	8361	83.61	0.8361	–	–	–
1 square mile	–	–	–	1	–	–	–	–	259	2.59
1 cm^2	0.155	–	–	–	1	0.01	–	–	–	–
1 dm^2	15.5	0.1076	0.01196	–	100	1	0.01	–	–	–
1 m^2	1550	10.76	1.196	–	10 000	100	1	0.01	–	–
1 are (a)	–	1076	119.6	–	–	10 000	100	1	0.01	–
1 hectare (ha)	–	–	–	–	–	–	10 000	100	1	0.01
1 km^2	–	–	–	0.3861	–	–	–	10 000	100	1

2.3.14 Length and area

Comparative lengths in USCS and SI units are:

1 ft = 0.3048 m
1 in = 25.4 mm
1 statute mile = 1609.3 m
1 nautical mile = 1853.2 m

The basic unit of area is square feet (ft^2) or square inches (in^2 or sq.in). In SI it is m^2. See Table 2.9.

Small dimensions are measured in 'micro-measurements' (see Figure 2.8).

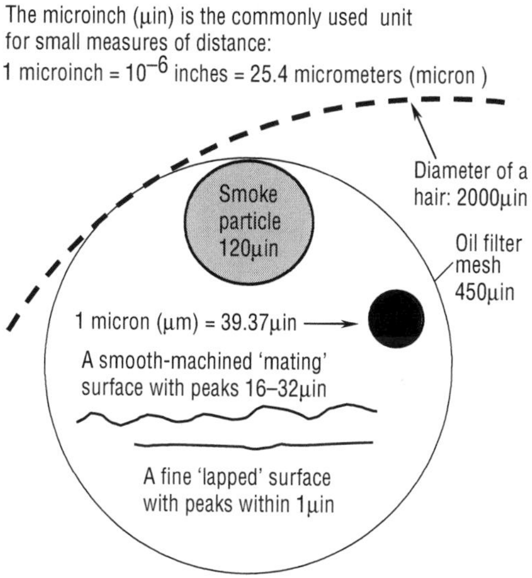

Fig. 2.8 Micromeasurements

2.3.15 Viscosity

Dynamic viscosity (μ) is measured in lbf.s/ft^2 or, in the SI system, in N s/m^2 or pascal seconds (Pa s).

1 lbf.s/ft^2 = 4.882 kgf.s/m^2 = 4.882 Pa s
1 Pa s = 1 N s/m^2 = 1 kg/m s

A common unit of viscosity is the centipoise (cP). See Table 2.10.

Table 2.10 Dynamic viscosity (μ)

Unit	*lbf-s/ft²*	*Centipoise*	*Poise*	*kgf/m s*
1 lb (force)-s per ft²	1	4.788×10^{4}	4.788×10^{2}	4.882
1 centipoise	2.089×10^{-5}	1	10^{-2}	1.020×10^{-4}
1 poise	2.089×10^{-3}	100	1	1.020×10^{-2}
1 N-s per m²	0.2048	9.807×10^{3}	98.07	1

Kinematic viscosity (ν) is a function of dynamic viscosity.

Kinematic viscosity = dynamic viscosity/density, i.e. $\nu = \mu/\rho$

The basic unit is ft²/s. Other units such as Saybolt Seconds Universal (SSU) are also used.

$1\ m^2/s = 10.7639\ ft^2/s = 5.58001 \times 10^6\ in^2/h$
1 stoke (St) $= 100$ centistokes (cSt) $= 10^{-4}\ m^2/s$
$1\ St >\cong 0.00226\ (SSU) - 1.95/(SSU)$ for $32 < SSU < 100$ seconds
$1\ St \cong 0.00220\ (SSU) - 1.35/(SSU)$ for $SSU > 100$ seconds

2.4 Consistency of units

Within any system of units, the consistency of units forms a 'quick check' of the validity of equations. The units must match on both sides.

Example:

To check kinematic viscosity (ν) =

$$\frac{\text{dynamic viscosity } (\mu)}{\text{density } (\rho)} = \mu \times 1/\rho$$

$$\frac{ft^2}{s} = \frac{lbf.s}{ft^2} \times \frac{ft^4}{lbf.s^2}$$

Cancelling gives $\dfrac{ft^2}{s} = \dfrac{s.ft^4}{s^2.ft^2} = \dfrac{ft^2}{s}$

OK, units match.

2.5 Foolproof conversions: using unity brackets

When converting between units it is easy to make mistakes by dividing by a conversion factor instead of multiplying, or vice versa. The best way to avoid this is by using the technique of unity brackets.

A unity bracket is a term, consisting of a numerator and denominator in different units, which has a value of unity.

e.g. $\left[\dfrac{2.205\ \text{lb}}{\text{kg}}\right]$ or $\left[\dfrac{\text{kg}}{2.205\ \text{lb}}\right]$ are unity brackets

as are

$$\left[\frac{25.4\ \text{mm}}{\text{in}}\right] \text{ or } \left[\frac{\text{in}}{25.4\ \text{mm}}\right] \text{ or } \left[\frac{\text{atmosphere}}{101\,325\ \text{Pa}}\right]$$

Remember that, as the value of the term inside the bracket is unity, it has no effect on any term that it multiplies.

Example:

Convert the density of titanium 6 Al 4 V; $\rho = 0.16\ \text{lb/in}^3$ to kg/m^3

Step 1: State the initial value: $\rho = \dfrac{0.16\ \text{lb}}{\text{in}^3}$

Step 2: Apply the 'weight' unity bracket:

$$\rho = \frac{0.16\ \text{lb}}{\text{in}^3}\left[\frac{\text{kg}}{2.205\ \text{lb}}\right]$$

Step 3: Then apply the 'dimension' unity brackets (cubed):

$$\rho = \frac{0.16\ \text{lb}}{\text{in}^3}\left[\frac{\text{kg}}{2.205\ \text{lb}}\right]^3\left[\frac{\text{in}}{25.4\ \text{mm}}\right]^3 \left[\frac{1000\ \text{mm}}{\text{m}}\right]^3$$

Step 4: Expand and cancel*:

$$\rho = \frac{0.16\,\text{lb}}{\text{in}^3}\left[\frac{\text{kg}}{2.205\,\text{lb}}\right]\left[\frac{\text{in}^3}{(25.4)^3\,\text{mm}^3}\right]\left[\frac{(1000)^3\,\text{mm}^3}{\text{m}^3}\right]$$

$$\rho = \frac{0.16\text{ kg }(1000)^3}{2.205\ (25.4)^3\text{ m}^3}$$

$\rho = 4428.02\text{ kg/m}^3$ Answer

*Take care to use the correct algebraic rules for the expansion, e.g.

$(a.b)^N = a^N.b^N$ not $a.b^N$

e.g. $\left[\dfrac{1000\text{ mm}}{\text{m}}\right]^3$ expands to $\dfrac{(1000)^3\ (\text{mm})^3}{(\text{m})^3}$

Unity brackets can be used for all unit conversions provided you follow the rules for algebra correctly.

2.6 Imperial–metric conversions

See Table 2.11.

2.7 Dimensional analysis

2.7.1 Dimensional analysis (DA) – what is it?

DA is a technique based on the idea that one physical quantity is related to others in a precise mathematical way.

It is used in aeronautics for:

- Checking the validity of equations.
- Finding the arrangement of variables in a formula.
- Helping to tackle problems that do not possess a compete theoretical solution – particularly those involving fluid mechanics.

2.7.2 Primary and secondary quantities

Primary quantities are quantities which are absolutely independent of each other. They are:

Table 2.11 Imperial-metric conversions

Fraction (in)	*Decimal (in)*	*Millimetre (mm)*	*Fraction (in)*	*Decimal (in)*	*Millimetre (mm)*
1/64	0.01562	0.39687	33/64	0.51562	13.09687
1/32	0.03125	0.79375	17/32	0.53125	13.49375
3/64	0.04687	1.19062	35/64	0.54687	13.89062
1/16	0.06250	1.58750	9/16	0.56250	14.28750
5/64	0.07812	1.98437	37/64	0.57812	14.68437
3/32	0.09375	2.38125	19/32	0.59375	15.08125
7/64	0.10937	2.77812	39/64	0.60937	15.47812
1/8	0.12500	3.17500	5/8	0.62500	15.87500
9/64	0.14062	3.57187	41/64	0.64062	16.27187
5/32	0.15625	3.96875	21/32	0.65625	16.66875
11/64	0.17187	4.36562	43/64	0.67187	17.06562
3/16	0.18750	4.76250	11/16	0.68750	17.46250
13/64	0.20312	5.15937	45/64	0.70312	17.85937
7/32	0.21875	5.55625	23/32	0.71875	18.25625
15/64	0.23437	5.95312	47/64	0.73437	18.65312
1/4	0.25000	6.35000	3/4	0.75000	19.05000
17/64	0.26562	6.74687	49/64	0.76562	19.44687
9/32	0.28125	7.14375	25/32	0.78125	19.84375
19/64	0.29687	5.54062	51/64	0.79687	20.24062
15/16	0.31250	7.93750	13/16	0.81250	20.63750
21/64	0.32812	8.33437	53/64	0.82812	21.03437
11/32	0.34375	8.73125	27/32	0.84375	21.43125
23/64	0.35937	9.12812	55/64	0.85937	21.82812
3/8	0.37500	9.52500	7/8	0.87500	22.22500
25/64	0.39062	9.92187	57/64	0.89062	22.62187
13/32	0.40625	10.31875	29/32	0.90625	23.01875
27/64	0.42187	10.71562	59/64	0.92187	23.41562
7/16	0.43750	11.11250	15/16	0.93750	23.81250
29/64	0.45312	11.50937	61/64	0.95312	24.20937
15/32	0.46875	11.90625	31/12	0.96875	24.60625
31/64	0.48437	12.30312	63/64	0.98437	25.00312
1/2	0.50000	12.70000	1	1.00000	25.40000

M Mass
L Length
T Time

For example, velocity (v) is represented by length divided by time, and this is shown by:

$[v] = \frac{L}{T}$: note the square brackets denoting 'the dimension of'.

Table 2.12 shows the most commonly used quantities.

Table 2.12 Dimensional analysis quantities

Quantity	*Dimensions*
Mass (m)	M
Length (l)	L
Time (t)	T
Area (a)	L^2
Volume (V)	L^3
First moment of area	L^3
Second moment of area	L^4
Velocity (v)	LT^{-1}
Acceleration (a)	LT^{-2}
Angular velocity (ω)	T^{-1}
Angular acceleration (α)	T^{-2}
Frequency (f)	T^{-1}
Force (F)	MLT^{-2}
Stress {pressure}, ($S\{P\}$)	$ML^{-1}T^{-2}$
Torque (T)	ML^2T^{-2}
Modulus of elasticity (E)	$ML^{-1}T^{-2}$
Work (W)	ML^2T^{-2}
Power (P)	ML^2T^{-3}
Density (ρ)	ML^{-3}
Dynamic viscosity (μ)	$ML^{-1}T^{-1}$
Kinematic viscosity (ξ)	L^2T^{-1}

Hence velocity is called a secondary quantity because it can be expressed in terms of primary quantities.

2.7.3 An example of deriving formulae using DA

To find the frequencies (n) of eddies behind a cylinder situated in a free stream of fluid, we can assume that n is related in some way to the diameter (d) of the cylinder, the speed (V) of the fluid stream, the fluid density (ρ) and the kinematic viscosity (ν) of the fluid.

i.e. $n = \phi\{d,V,\rho,\nu\}$

Introducing a numerical constant Y and some possible exponentials gives:

$$n = Y\{d^a,V^b,\rho^c,\nu^d\}$$

Y is a dimensionless constant so, in dimensional analysis terms, this equation becomes, after substituting primary dimensions:

$$T^{-1} = L^a(LT^{-1})^b\,(ML^{-3})^c\,(L^2T^{-1})^d$$
$$= L^a\,L^b\,T^{-b}\,M^c\,L^{-3c}\,L^{2d}\,T^{-d}$$

In order for the equation to balance:

For M, c must $= 0$
For L, $a + b - 3c + 2d = 0$
For T, $-b - d = -1$

Solving for a, b, c in terms of d gives:

$a = -1 - d$
$b = 1 - d$

Giving

$$n = d^{(-1-d)}\,V^{(1-d)}\,\rho^0\,\nu^d$$

Rearranging gives:

$$nd/V = (Vd/\nu)X$$

Note how dimensional analysis can give the 'form' of the formula but not the numerical value of the undetermined constant X which, in this case, is a compound constant containing the original constant Y and the unknown index d.

2.8 Essential mathematics

2.8.1 Basic algebra

$a^m \times a^n = a^{m+n}$

$a^m \div a^n = a^{m-n}$

$(a^m)^n = a^{mn}$

$\sqrt[n]{a^m} = a^{m/n}$

$\dfrac{1}{a^n} = a^{-n}$

$a^o = 1$

$(a^n b^m)^p = a^{np}\,b^{mp}$

$\left(\dfrac{a}{b}\right)^n = \dfrac{a^n}{b^n}$

$\sqrt[n]{ab} = \sqrt[n]{a} \times \sqrt[n]{b}$

$\sqrt[n3]{a\backslash b} = \dfrac{\sqrt[n]{a}}{\sqrt[n]{b}}$

2.8.2 Logarithms

If $N = a^x$ then $\log_a N = x$ and $N = a^{\log_a N}$

$$\log_a N = \frac{\log_b N}{\log_b a}$$

$$\log(ab) = \log a + \log b$$

$$\log\left(\frac{a}{b}\right) = \log a - \log b$$

$$\log a^n = n \log a$$

$$\log \sqrt[n]{a} = \frac{1}{n} \log a$$

$$\log_a 1 = 0$$
$$\log_e N = 2.3026 \log_{10} N$$

2.8.3 Quadratic equations

If $ax^2 + bx + c = 0$

$$x = \frac{-b \pm \sqrt{b^2 - 4ac}}{2a}$$

If $b^2 - 4ac > 0$ the equation $ax^2 + bx + c = 0$ yields two real and different roots.

If $b^2 - 4ac = 0$ the equation $ax^2 + bx + c = 0$ yields coincident roots.

If $b^2 - 4ac < 0$ the equation $ax^2 + bx + c = 0$ has complex roots.

If α and β are the roots of the equation $ax^2 + bx + c = 0$ then

$$\text{sum of the roots} = \alpha + \beta = -\frac{b}{a}$$

$$\text{product of the roots} = \alpha\beta = \frac{c}{d}$$

The equation whose roots are α and β is $x^2 - (\alpha + \beta)x + \alpha\beta = 0$.

Any quadratic function $ax^2 + bx + c$ can be expressed in the form $p\,(x + q)^2 + r$ or $r - p\,(x + q)^2$, where r, p and q are all constants.

The function $ax^2 + bx + c$ will have a maximum value if a is negative and a minimum value if a is positive.

If $ax^2 + bx + c = p(x + q)^2 + r = 0$ the minimum value of the function occurs when $(x + q) = 0$ and its value is r.
If $ax^2 + bx + c = r - p(x + q)^2$ the maximum value of the function occurs when $(x + q) = 0$ and its value is r.

2.8.4 Cubic equations

$$x^3 + px^2 + qx + r = 0$$
$$x = y - \tfrac{1}{3}p \quad \text{gives} \quad y^3 + 3ay + 2b = 0$$

where

$$3a = -q - \tfrac{1}{3}p^2, \quad 2b = \tfrac{2}{27}p^3 - \tfrac{1}{3}pq + r$$

On setting

$$S = [-b + (b^2 + a^3)^{1/2}]^{1/3}$$

and

$$T = [-b - (b^2 + a^3)^{1/2}]^{1/3}$$

the three roots are

$$x_1 = S + T - \tfrac{1}{3}p$$

$$x_2 = -\tfrac{1}{2}(S + T) + \sqrt{3}\backslash 2\, i(S - T) - \tfrac{1}{3}p$$

$$x_3 = -\tfrac{1}{2}(S + T) - \sqrt{3}\backslash 2\, i(S - T) - \tfrac{1}{3}p.$$

For real coefficients

all roots are real if $b^2 + a^3 \leq 0$,
one root is real if $b^2 + a^3 > 0$.

At least two roots are equal if $b^2 + a^3 = 0$. Three roots are equal if $a = 0$ and $b = 0$. For $b^2 + a^3 < 0$
there are alternative expressions:

$$x_1 = 2c \cos\tfrac{1}{3}\theta - \tfrac{1}{3}p \quad x_2 = 2c \cos\tfrac{1}{3}(\theta + 2\pi) - \tfrac{1}{3}p$$
$$x_3 = 2c \cos\tfrac{1}{3}(\theta + 4\pi) - \tfrac{1}{3}p$$

where $c^2 = -a$ and $\cos\theta = -\dfrac{b}{c^3}$

2.8.5 Complex numbers

If x and y are real numbers and $i = \sqrt{-1}$ then the complex number $z = x + iy$ consists of the real part x and the imaginary part iy.
$\bar{z} = x - iy$ is the conjugate of the complex number $z = x + iy$.

If $x + iy = a + ib$ then $x = a$ and $y = b$

$$(a + ib) + (c + id) = (a + c) = i(b + d)$$
$$(a + ib) - (c + id) = (a - c) = i(b + d)$$
$$(a + ib)(c + id) = (ac - bd) + i(ad + bc)$$

$$\frac{a + ib}{c + id} = \frac{ac + bd}{c^2 + d^2} + i\,\frac{bc - ad}{c^2 + d^2}$$

Every complex number may be written in polar form. Thus

$$x + iy = r(\cos\theta + i \sin \theta) = r\angle\theta$$

r is called the modulus of z and this may be written $r = |z|$

$$r = \sqrt{x^2 + y^2}$$

θ is called the argument and this may be written $\theta = \arg z$

$$\tan \theta = \frac{y}{x}$$

If $z_1 = r\ (\cos\theta_1 + i \sin \theta_1)$ and $z_2 = r_2\ (\cos\theta_2 + i \sin \theta_2)$

$$z_1 z_2 = r_1 r_2\ [\cos(\theta_1 + \theta_2) + i \sin(\theta_1 + \theta_2)]$$
$$= r_1 r_2 \angle(\theta_1 + \theta_2)$$

$$z_1 \backslash z_2 = \frac{r_1[\cos(\theta_1 - \theta_2) + i \sin(\theta_1 + \theta_2)]}{r_2}$$

$$= \frac{r_1}{r_2} \angle(\theta_1 - \theta_2)$$

2.8.6 Standard series

Binomial series

$$(a + x)^n = a^n + na^{n-1}\,x + \frac{n(n - 1)}{2!}\,a^{n-2}\,x^2$$
$$+ \frac{n(n - 1)(n - 2)}{3!}\,a^n{-}3\,x^3$$
$$+ \ldots\ (x^2 < a^2)$$

The number of terms becomes inifinite when n is negative or fractional.

$$(a - bx)^{-1} = \frac{1}{a}\left(1 + \frac{bx}{a} + \frac{b^2x^2}{a^2} + \frac{b^3x^3}{a^3} + \ldots\right)$$
$$(b^2 x^2 < a^2)$$

Exponential series

$$a^x = 1 + x \ln a + \frac{(x \ln a)^2}{2!} + \frac{(x \ln a)^3}{3!} + \ldots$$

$$e^x = 1 + x + \frac{x^2}{2!} + \frac{x^3}{3!} + \ldots$$

Logarithmic series

$$\ln x = (x - 1) - \tfrac{1}{2}(x - 1)^2 + \tfrac{1}{3}(x - 1)^3 - \ldots (0 < x < 2)$$

$$\ln x = \frac{x - 1}{x} + \tfrac{1}{2}\left(\frac{x - 1}{x}\right)^2 + \tfrac{1}{3}\left(\frac{x - 1}{x}\right)^3 + \ldots \left(x > \frac{1}{2}\right)$$

$$\ln x = 2[\frac{x - 1}{x + 1} \cdot \frac{1}{3}\left(\frac{x - 1}{x + 1}\right)^3 + \frac{1}{5}\left(\frac{x - 1}{x + 1}\right)^5 + \ldots (x \text{ positive})$$

$$\ln (1 + x) = x - \frac{x^2}{2} + \frac{x^3}{3} - \frac{x^4}{4} + \ldots$$

Trigonometric series

$$\sin x = x - \frac{x^3}{3!} + \frac{x^5}{5!} - \frac{x^7}{7!} + \ldots$$

$$\cos x = 1 - \frac{x^2}{2!} + \frac{x^4}{4!} - \frac{x^6}{6!} + \ldots$$

$$\tan x = x + \frac{x^3}{3} + \frac{2x^5}{15} + \frac{17x^7}{315} + \frac{62x^9}{2835} + \ldots \left(x^2 < \frac{\pi^2}{4}\right)$$

$$\sin^{-1} x = x + \frac{1}{2}\frac{x^3}{3} + \frac{1\cdot 3}{2\cdot 4} + \frac{x^5}{5} + \frac{1\cdot 3\cdot 5}{2\cdot 4\cdot 6}\frac{x^7}{7} + \ldots (x^2 < 1)$$

$$\tan^{-1} x = x - \frac{1}{3}x^3 + \frac{1}{5}x^5 - \frac{1}{7}x^7 + \ldots \ (x^2 \leqq 1)$$

2.8.7 Vector algebra

Vectors have direction and magnitude and satisfy the *triangle rule* for addition. Quantities such as velocity, force, and straight-line displacements may be represented by vectors. Three-dimensional vectors are used to represent physical quantities in space, e.g. A_x, A_y, A_z or $A_x\mathbf{i} + A_y\mathbf{j} + A_z\mathbf{k}$.

Vector Addition

The vector sum $\mathbf{V}$ of any number of vectors $\mathbf{V}_1$, $\mathbf{V}_2$, $\mathbf{V}_3$ where $= \mathbf{V}_1\ a_1\mathbf{i} + b_1\,\mathbf{j} + c_1\,\mathbf{k}$, etc., is given by

$$\mathbf{V} = \mathbf{V}_1 + \mathbf{V}_2 + \mathbf{V}_3 + \ldots = (a_1 + a_2 + a_3 + \ldots)\mathbf{i}$$
$$+(b_1 + b_2 + b_3 + \ldots)\mathbf{j} + (c_1 + c_2 + c_3 + \ldots)\mathbf{k}$$

Product of a vector $\mathbf{V}$ *by a scalar quantity s*

$$\mathbf{s}\mathbf{V} = (sa)\mathbf{i} + (sb)\mathbf{j} + (sc)\mathbf{k}$$
$$(\mathbf{s}_1 + \mathbf{s}_2)\mathbf{V} = \mathbf{s}_1\mathbf{V} + \mathbf{s}_2\mathbf{V} \quad (\mathbf{V}_1 + \mathbf{V}_2)\mathbf{s} = \mathbf{V}_1\mathbf{s} + \mathbf{V}_2\mathbf{s}$$

where $\mathbf{sV}$ has the same direction as $\mathbf{V}$, and its magnitude is $\mathbf{s}$ times the magnitude of $\mathbf{V}$.

Scalar product of two vectors, $\mathbf{V}_1 \cdot \mathbf{V}_2$

$$\mathbf{V}_1 \cdot \mathbf{V}_2 = |\mathbf{V}_1||\mathbf{V}_2|\cos\phi$$

Vector product of two vectors, $\mathbf{V}_1 \times \mathbf{V}_2$

$$\mathbf{V}_1 \times \mathbf{V}_2| = |\mathbf{V}_1||\mathbf{V}_2|\sin\phi$$

where ϕ is the angle between $\mathbf{V}_1$ and $\mathbf{V}_2$.

Derivatives of vectors

$$\frac{d}{dt}(\mathbf{A} \cdot \mathbf{B}) = \mathbf{A} \cdot \frac{d\mathbf{B}}{dt} + \mathbf{B} \cdot \frac{d\mathbf{A}}{dt}$$

If $e(t)$ is a unit vector $\frac{de}{dt}$ is perpendicular to e: that is $e \cdot \frac{de}{dt} = 0$.

$$\frac{d}{dt}(\mathbf{A} \times \mathbf{B}) = \mathbf{A} \times \frac{d\mathbf{B}}{dt} + \frac{d\mathbf{A}}{dt} \times \mathbf{B}$$

$$= -\frac{d}{dt}(\mathbf{B} \times \mathbf{A})$$

Gradient

The gradient (grad) of a scalar field $\phi(x, y, z)$ is

$$\text{grad } \phi = \nabla\phi = \left(\mathbf{i}\frac{\partial}{\partial x} + \mathbf{j}\frac{\partial}{\partial y} + \mathbf{k}\frac{\partial}{\partial z}\right)\phi$$

$$= \frac{\partial\phi}{\partial x}\mathbf{i} + \frac{\partial\phi}{\partial y}\mathbf{j}\frac{\partial\phi}{\partial z}\mathbf{k}$$

Divergence

The divergence (div) of a vector $\mathbf{V} = \mathbf{V}(x, y, z) = V_x(x, y, z)\,\mathbf{i} + V_y\,(x, y, z)\,\mathbf{j} + \mathrm{V}_z\,(x, y, z)\mathbf{k}$

$$\text{div } \mathbf{V} = \nabla \cdot \mathbf{V}\frac{\partial \mathbf{V}_x}{\partial x} + \frac{\partial \mathbf{V}_y}{\partial y} + \frac{\partial \mathbf{V}_z}{\partial z}$$

Curl

Curl (rotation) is:

$$\text{curl } \mathbf{V} = \nabla \times \mathbf{V} = \begin{matrix} \mathbf{i} & \mathbf{j} & \mathbf{k} \\ \frac{\partial}{\partial x} & \frac{\partial}{\partial y} & \frac{\partial}{\partial z} \\ \mathrm{V}_x & \mathrm{V}_y & \mathrm{V}_z \end{matrix}$$

$$= \left(\frac{\partial \mathbf{V}_z}{\partial y} - \frac{\partial \mathbf{V}_y}{\partial z}\right)\mathbf{i} + \left(\frac{\partial \mathbf{V}_x}{\partial z} - \frac{\partial \mathbf{V}_z}{\partial x}\right)\mathbf{j}$$

$$+ \left(\frac{\partial \mathbf{V}_y}{\partial x} - \frac{\partial \mathbf{V}_x}{\partial y}\right)\mathbf{k}$$

2.8.8 Differentiation

Rules for differentiation: y, u and v are functions of x; a, b, c and n are constants.

$$\frac{d}{dx}(au \pm bv) = a\frac{du}{dx} \pm b\frac{dv}{dx}$$

$$\frac{d\,(uv)}{dx} = u\frac{dv}{dx} + v\frac{du}{dx}$$

$$\frac{d}{dx}\left(\frac{u}{v}\right) = \frac{1}{v}\frac{du}{dx} - \frac{u}{v^2}\frac{dv}{dx}$$

$$\frac{d}{dx}(u^n) = nu^{n-1}\frac{du}{dx}, \quad \frac{d}{dx}\left(\frac{1}{u^n}\right) = -\frac{n}{u^{n+1}}\frac{du}{dx}$$

$$\frac{du}{dx} = 1\bigg/\frac{dx}{du}, \quad \text{if } \frac{dx}{du} \neq 0$$

$$\frac{d}{dx}f(u) = f'(u)\frac{du}{dx}$$

$$\frac{d}{dx}\int_a^x f(t)dt = f(x)$$

$$\frac{d}{dx}\int_x^b f(t)dt = -f(x)$$

$$\frac{d}{dx}\int_a^b f(x, t)dt = \int_a^b \frac{\partial f}{\partial x}dt$$

$$\frac{d}{dx}\int_u^v f(x, t)dt = \int_v^u \frac{\partial f}{\partial x}dt + f(x, v)\frac{dv}{dx} - f(x, u)\frac{du}{dx}$$

Higher derivatives

$$\text{Second derivatives} = \frac{d}{dx}\left(\frac{dy}{dx}\right) = \frac{d^2y}{dx^2} = f''(x) = y''$$

$$\frac{d^2}{dx^2}f(u) = f''(u)\left(\frac{du}{dx}\right)^2 + f'(u)\frac{d^2u}{dx^2}$$

Derivatives of exponentials and logarithms

$$\frac{d}{dx}(ax + b)^n = na(ax + b)^{n-1}$$

$$\frac{d}{dx}e^{ax} = ae^{ax}$$

$$\frac{d}{dx}\ln ax = \frac{1}{x}, \quad ax > 0$$

$$\frac{d}{dx} a^u = a^u \ln a \frac{du}{dx}$$

$$\frac{d}{dx} \log_a u = \log_a e \frac{1}{u} \frac{du}{dx}$$

Derivatives of trigonometric functions in radians

$$\frac{d}{dx} \sin x = \cos x, \quad \frac{d}{dx} \cos x = -\sin x$$

$$\frac{d}{dx} \tan x = \sec^2 x = 1 + \tan^2 x$$

$$\frac{d}{dx} \cot x = -\mathrm{cosec}^2 x$$

$$\frac{d}{dx} \sec x = \frac{\sin x}{\cos^2 x} = \sec x \tan x$$

$$\frac{d}{dx} \mathrm{cosec}\, x = -\frac{\cos x}{\sin^2 x} = -\mathrm{cosec}\, x \cot x$$

$$\frac{d}{dx} \arcsin x = -\frac{d}{dx} \arccos x$$

$$= \frac{1}{(1 - x^2)^{1/2}}$$ for angles in the first quadrant.

Derivatives of hyperbolic functions

$$\frac{d}{dx} \sinh x = \cosh x, \quad \frac{d}{dx} \cosh x = \sinh x$$

$$\frac{d}{dx} \tanh x = \mathrm{sech}^2 x, \quad \frac{d}{dx} \cosh x = -\mathrm{cosech}^2 x$$

$$\frac{d}{dx} (\mathrm{arcsinh}\, x) = \frac{1}{(x^2 + 1)^{1/2}},$$

$$\frac{d}{dx} (\mathrm{arccosh}\, x) = \frac{\pm 1}{(x^2 - 1)^{1/2}}$$

Partial derivatives Let $f(x, y)$ be a function of the two variables x and y. The partial derivative of f with respect to x, keeping y constant is:

$$\frac{\partial f}{\partial x} = \lim_{h \to 0} \frac{f(x+h, y) - f(x, y)}{h}$$

Similarly the partial derivative of f with respect to y, keeping x constant, is

$$\frac{\partial f}{\partial y} = \lim_{k \to 0} \frac{f(x, y+k) - f(x, y)}{k}$$

Chain rule for partial derivatives To change variables from (x, y) to (u, v) where $u = u(x, y)$, $v = v(x, y)$, both $x = x(u, v)$ and $y(u, v)$ exist and $f(x, y) = f[x(u, v), y(u, v)] = F(u, v)$.

$$\frac{\partial F}{\partial u} = \frac{\partial x}{\partial u}\frac{\partial f}{\partial x} + \frac{\partial y}{\partial u}\frac{\partial f}{\partial y}, \quad \frac{\partial F}{\partial v} = \frac{\partial x}{\partial v}\frac{\partial f}{\partial v} + \frac{\partial y}{\partial v}\frac{\partial f}{\partial y}$$

$$\frac{\partial f}{\partial x} = \frac{\partial u}{\partial x}\frac{\partial F}{\partial u} + \frac{\partial v}{\partial x}\frac{\partial F}{\partial v}, \quad \frac{\partial f}{\partial y} = \frac{\partial u}{\partial y}\frac{\partial F}{\partial u} + \frac{\partial v}{\partial y}\frac{\partial F}{\partial v}$$

2.8.9 Integration

$f(x)$	$F(x) = \int f(x)dx$
x^a	$\frac{x^{a+1}}{a+1}, \quad a \neq -1$
x^{-1}	$\ln \mid x \mid$
e^{kx}	$\frac{e^{kx}}{k}$
a^x	$\frac{a^x}{\ln a}, \quad a > 0, \quad a \neq 1$
$\ln x$	$x \ln x - x$
$\sin x$	$-\cos x$
$\cos x$	$\sin x$
$\tan x$	$\ln \mid \sec x \mid$
$\cot x$	$\ln \mid \sin x \mid$
$\sec x$	$\ln \mid \sec x + \tan x \mid$ $= \ln \mid \tan \frac{1}{2}(x + \frac{1}{2}\pi) \mid$

$\operatorname{cosec} x$	$\ln \lvert \tan \frac{1}{2} x \rvert$
$\sin^2 x$	$\frac{1}{2}(x - \frac{1}{2}\sin 2x)$
$\cos^2 x$	$\frac{1}{2}(x + \frac{1}{2}\sin 2x)$
$\sec^2 x$	$\tan x$
$\sinh x$	$\cosh x$
$\cosh x$	$\sinh x$
$\tanh x$	$\ln \cosh x$
$\operatorname{sech} x$	$2 \arctan e^x$
$\operatorname{cosech} x$	$\ln \lvert \tanh \frac{1}{2} x \rvert$
$\operatorname{sech}^2 x$	$\tanh x$
$\dfrac{1}{a^2 + x^2}$	$\dfrac{1}{a} \arctan \dfrac{x}{a}, \quad a \neq 0$
$\dfrac{1}{a^2 - x^2}$	$\begin{cases} -\dfrac{1}{2a} \ln \dfrac{a - x}{a + x}, & a \neq a \\ \dfrac{1}{2a} \ln \dfrac{x - a}{x + a}, & a \neq 0 \end{cases}$
$\dfrac{1}{(a^2 - x^2)^{1/2}}$	$\arcsin \dfrac{x}{\lvert a \rvert}, \quad a \neq 0$
$\dfrac{1}{(x^2 - a^2)^{1/2}}$	$\begin{cases} \ln [x + (x^2 - a^2)^{1/2}] \\ \operatorname{arccosh} \dfrac{x}{a}, \quad a \neq 0 \end{cases}$

2.8.10 Matrices

A matrix which has an array of $m \times n$ numbers arranged in m rows and n columns is called an $m \times n$ matrix. It is denoted by:

$$\begin{bmatrix} a_{11} & a_{12} & \cdots & a_{1n} \\ a_{21} & a_{22} & \cdots & a_{2n} \\ . & . & \cdots & . \\ . & . & \cdots & . \\ . & . & \cdots & . \\ a_{m1} & a_{m2} & \cdots & a_{mn} \end{bmatrix}$$

Square matrix
This is a matrix having the same number of rows and columns.

$$\begin{bmatrix} a_{11} & a_{12} & a_{13} \\ a_{21} & a_{22} & a_{23} \\ a_{31} & a_{32} & a_{33} \end{bmatrix}$$ is a square matrix of order 3 × 3.

Diagonal matrix
This is a square matrix in which all the elements are zero except those in the leading diagonal.

$$\begin{bmatrix} a_{11} & 0 & 0 \\ 0 & a_{22} & 0 \\ 0 & 0 & a_{33} \end{bmatrix}$$ is a diagonal matrix of order 3 × 3.

Unit matrix
This is a diagonal matrix with the elements in the leading diagonal all equal to 1. All other elements are 0. The unit matrix is denoted by *I*.

$$I = \begin{bmatrix} 1 & 0 & 0 \\ 0 & 1 & 0 \\ 0 & 0 & 1 \end{bmatrix}$$

Addition of matrices
Two matrices may be added provided that they are of the same order. This is done by adding the corresponding elements in each matrix.

$$\begin{bmatrix} a_{11} & a_{12} & a_{13} \\ a_{21} & a_{22} & a_{23} \end{bmatrix} + \begin{bmatrix} b_{11} & b_{12} & b_{13} \\ b_{21} & b_{22} & b_{23} \end{bmatrix}$$

$$= \begin{bmatrix} a_{11} + b_{11} & a_{12} + b_{12} & a_{13} + b_{13} \\ a_{21} + b_{21} & a_{22} + b_{22} & a_{23} + b_{23} \end{bmatrix}$$

Subtraction of matrices
Subtraction is done in a similar way to addition except that the corresponding elements are subtracted.

$$\begin{bmatrix} a_{11} & a_{12} \\ a_{21} & a_{22} \end{bmatrix} - \begin{bmatrix} b_{11} & b_{12} \\ b_{21} & b_{22} \end{bmatrix} = \begin{bmatrix} a_{11} - b_{11} & a_{12} - b_{12} \\ a_{21} - b_{21} & a_{22} - b_{22} \end{bmatrix}$$

Scalar multiplication

A matrix may be multiplied by a number as follows:

$$b\begin{bmatrix} a_{11} & a_{12} \\ a_{21} & a_{22} \end{bmatrix} = \begin{bmatrix} ba_{11} & ba_{12} \\ ba_{21} & ba_{22} \end{bmatrix}$$

General matrix multiplication

Two matrices can be multiplied together provided the number of columns in the first matrix is equal to the number of rows in the second matrix.

$$\begin{bmatrix} a_{11} & a_{12} & a_{13} \\ a_{21} & a_{22} & a_{23} \end{bmatrix}\begin{bmatrix} b_{11} & b_{12} \\ b_{21} & b_{22} \\ b_{31} & b_{32} \end{bmatrix}$$

$$= \begin{bmatrix} a_{11}b_{11} + a_{12}b_{22} + a_{13}b_{31} & a_{11}b_{12} + a_{12}b_{22} + a_{13}b_{32} \\ a_{21}b_{11} + a_{22}b_{21} + a_{23}b_{31} & a_{21}b_{12} + a_{22}b_{22} + a_{23}b_{32} \end{bmatrix}$$

If matrix A is of order $(p \times q)$ and matrix B is of order $(q \times r)$ then if $C = AB$, the order of C is $(p \times r)$.

Transposition of a matrix

When the rows of a matrix are interchanged with its columns the matrix is said to be *transposed*. If the original matrix is denoted by A, its transpose is denoted by A' or A^T.

$$\text{If } A = \begin{bmatrix} a_{11} & a_{12} & a_{13} \\ a_{21} & a_{22} & a_{23} \end{bmatrix} \text{ then } A^T = \begin{bmatrix} a_{11} & a_{21} \\ a_{12} & a_{22} \\ a_{13} & a_{23} \end{bmatrix}$$

Adjoint of a matrix

If $A = [a_{ij}]$ is any matrix and A_{ij} is the cofactor of a_{ij} the matrix $[A_{ij}]^T$ is called the *adjoint* of A. Thus:

$$A = \begin{bmatrix} a_{11} & a_{12} & \dots & a_{1n} \\ a_{21} & a_{22} & \dots & a_{2n} \\ \cdot & \cdot & & \cdot \\ \cdot & \cdot & & \cdot \\ \cdot & \cdot & & \cdot \\ a_{n1} & a_{n2} & \dots & a_{mn} \end{bmatrix} \quad \text{adj } A = \begin{bmatrix} A_{11} & A_{21} & \dots & A_{n1} \\ A_{12} & A_{22} & \dots & A_{n2} \\ \cdot & \cdot & & \cdot \\ \cdot & \cdot & & \cdot \\ \cdot & \cdot & & \cdot \\ A_{1n} & A_{2n} & \dots & A_{nn} \end{bmatrix}$$

Singular matrix
A square matrix is singular if the determinant of its coefficients is zero.

The inverse of a matrix
If A is a non-singular matrix of order $(n \times n)$ then its inverse is denoted by A^{-1} such that $AA^{-1} = I = A^{-1} A$.

$$A^{-1} = \frac{\text{adj }(A)}{\Delta} \quad \Delta = \det (A)$$

$$A_{ij} = \text{cofactor of } a_{ij}$$

$$\text{If } A = \begin{bmatrix} a_{11} & a_{12} & \dots & a_{1n} \\ a_{21} & a_{22} & \dots & a_{2n} \\ \cdot & \cdot & \dots & \cdot \\ \cdot & \cdot & \dots & \cdot \\ \cdot & \cdot & \dots & \cdot \\ a_{n1} & a_{n2} & \dots & a_{nn} \end{bmatrix} \quad A^{-1} = \frac{1}{\Delta} \begin{bmatrix} A_{11} & A_{21} & \dots & A_{n1} \\ A_{12} & A_{22} & \dots & A_{n2} \\ \cdot & & \dots & \cdot \\ \cdot & & \dots & \cdot \\ \cdot & & \dots & \cdot \\ A_{1n} & A_{2n} & \dots & A_{nn} \end{bmatrix}$$

2.8.11 Solutions of simultaneous linear equations

The set of linear equations

$$\begin{matrix} a_{11}x_1 + a_{12}x_2 + \dots + a_{1n}x_n = b_1 \\ a_{21}x_1 + a_{22}x_2 + \dots + a_{2n}x_n = b_2 \\ \vdots \qquad \vdots \qquad \qquad \vdots \qquad \vdots \\ a_{n1}x_1 + a_{n2}x_2 + \dots + a_{nn}x_n = b_n \end{matrix}$$

where the *a*s and *b*s are known, may be represented by the single matrix equation $Ax = b$, where A is the $(n \times n)$ matrix of coefficients, a_{ij}, and x and b are $(n \times 1)$ column vectors. The solution to this matrix equation, if A is non-singular, may be written as $x = A^{-1}b$ which leads to a solution given by *Cramer's rule*:

$$x_i = \det D_i / \det A \quad i = 1, 2, \dots, n$$

where $\det D_i$ is the determinant obtained from $\det A$ by replacing the elements of a_{ki} of the ith column by the elements b_k $(k = 1, 2, \dots, n)$. Note that this rule is obtained by using $A^{-1} = (\det A)^{-1}$ adj A and so again is of practical use only when $n \leq 4$.

If det $A = 0$ but det $D_i \neq 0$ for some i then the equations are inconsistent: for example, $x + y = 2$, $x + y = 3$ has no solution.

2.8.12 Ordinary differential equations

A *differential equation* is a relation between a function and its derivatives. The order of the highest derivative appearing is the *order* of the differential equation. Equations involving only one independent variable are *ordinary* differential equations, whereas those involving more than one are *partial* differential equations.

If the equation involves no products of the function with its derivatives or itself nor of derivatives with each other, then it is *linear*. Otherwise it is *non-linear*.

A linear differential equation of order n has the form:

$$P_0 \frac{d^n y}{dx^n} + P_1 \frac{d^{n-1} y}{dx^n{-}1} + \ldots + P_{n-1} \frac{dy}{dx} + P_n y = F$$

where P_i $(i = 0, 1. \ldots, n)$ F may be functions of x or constants, and $P_0 \neq 0$.

First order differential equations

Form	*Type*	*Method*
$\frac{dx}{dy} = f\left(\frac{y}{x}\right)$	homogeneous	substitute $u = \frac{y}{x}$
$\frac{dy}{dx} = f(x)g(y)$	separable	$\int \frac{dy}{g(y)} = \int f(x)dx + C$ note that roots of $g(y) = 0$ are also solutions
$g(x, y)\frac{dy}{dx} + f(x, y) = 0$ and $\frac{\partial f}{\partial y} = \frac{\partial g}{\partial x}$	exact	put $\frac{\partial \phi}{\partial x} = f$ and $\frac{\partial \phi}{\partial y} = g$ and solve these equations for ϕ; $\phi\,(x, y) = \text{constant}$ is the solution

$\frac{dy}{dx} + f(x)y = g(x)$	linear	Multiply through by $p(x) = \exp(\int^x f(t)dt)$ giving: $p(x)y = \int^x g(s)p(s)ds + C$

Second order (linear) equations
These are of the form:

$$P_0(x)\frac{d^2 y}{dx^2} + P_1(x)\frac{dy}{dx} + P_2(x)y = F(x)$$

When P_0, P_1, P_2 are constants and $f(x) = 0$, the solution is found from the roots of the auxiliary equation:

$$P_0 m_2 + P_1 m + P_2 = 0$$

There are three other cases:

(i) Roots $m = \alpha$ and β are real and $\alpha \neq \beta$

$$y(x) = Ae^{\alpha x} + Be^{\beta x}$$

(ii) Double roots: $\alpha = \beta$

$$y(x) = (A + Bx)e^{\alpha x}$$

(iii) Roots are complex: $m = k \pm il$

$$y(x) = (A \cos lx + \mathrm{B} \sin lx)e^{kx}$$

2.8.13 Laplace transforms
If $f(t)$ is defined for all t in $0 \leq t < \infty$, then

$$L[f(t)] = F(s) = \int_0^\infty e^{-st} f(t)dt$$

is called the Laplace transform of $f(t)$. The two functions of $f(t)$, $F(s)$ are known as a transform pair, and

$$f(t) = L^{-1}[F(s)]$$

is called the *inverse transform* of $F(s)$.

Function	*Transform*
$f(t), g(t)$	$F(s), G(s)$
$c_1 f(t) + c_2 g(t)$	$c_1 F(s) + c_2 G(s)$

$\int_0^t f(x)dx$	$F(s)/s$
$(-t)^n f(t)$	$\dfrac{d^n F}{ds^n}$
$e^{at} f(t)$	$F(s-a)$
$f(t-a)H(t-a)$	$e^{-as} F(s)$
$\dfrac{d^n f}{dt^n}$	$s^n F(s) - \sum_{r=1}^{n} s^{n-r} f^{(r-1)} (0+)$
$\dfrac{1}{a} e^{-bt} \sin at, \quad a > 0$	$\dfrac{1}{(s = b)^2 + a^2}$
$e^{-bt} \cos at$	$\dfrac{s+b}{(s+b)^2 + a^2}$
$\dfrac{1}{a} e^{-bt} \sinh at, \quad a > 0$	$\dfrac{1}{(s+b)^2 + a^2}$
$e^{-bt} \cosh at$	$\dfrac{s+b}{(s+b)^2 + a^2}$
$(\pi t)^{-1/2}$	$s^{-1/2}$
$\dfrac{2^n t^{n-1/2}}{1 \cdot 3 \cdot 5 \ldots (2n-1)\sqrt{\pi}}$, n integer	$s^{-(n+1/2)}$
$\dfrac{\exp(-a^2/4t)}{2(\pi t^3)^{1/2}} \quad (a > 0)$	$e^{-a\sqrt{s}}$

2.8.14 Basic trigonometry

Definitions (see Figure 2.9)

sine: $\sin A = \dfrac{y}{r}$ cosine: $\cos A = \dfrac{x}{r}$

tangent: $\tan A = \dfrac{y}{x}$ cotangent: $\cot A = \dfrac{x}{y}$

secant: $\sec A = \dfrac{r}{x}$ cosecant: $\text{cosec}\, A = \dfrac{r}{y}$

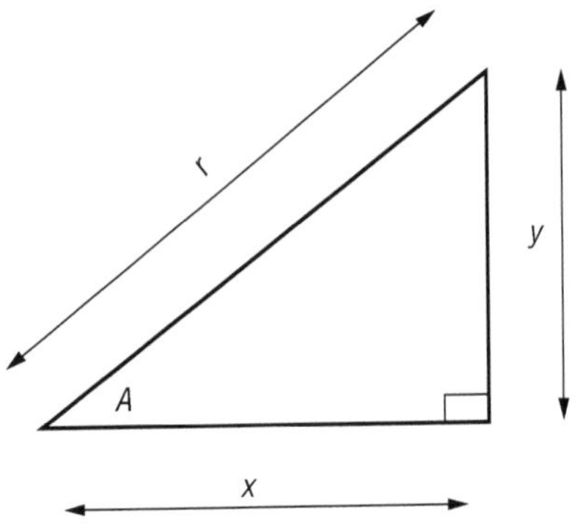

Fig. 2.9 Basic trigonometry

Relations between trigonometric functions

$$\sin^2 A + \cos^2 A = 1 \quad \sec^2 A = 1 + \tan^2 A$$
$$\operatorname{cosec}^2 A = 1 + \cot^2 A$$

$\sin A = s$	$\cos A = c$	$\tan A = t$	
$\sin A$	s	$(1 - c^2)^{1/2}$	$t(1 + t^2)^{-1/2}$
$\cos A$	$(1 - s^2)^{1/2}$	c	$(1 + t^2)^{-1/2}$
$\tan A$	$s(1 - s^2)^{1/2}$	$(1 - c^2)^{1/2}/c$	t

A is assumed to be in the first quadrant; signs of square roots must be chosen appropriately in other quadrants.

Addition formulae

$$\sin(A \pm B) = \sin A \cos B \pm \cos A \sin B$$
$$\cos(A \pm B) = \cos A \cos B \mp \sin A \sin B$$

$$\tan(A \pm B) = \frac{\tan A \pm \tan B}{1 \mp \tan A \tan B}$$

Sum and difference formulae

$$\sin A + \sin B = 2 \sin \tfrac{1}{2}(A + B) \cos \tfrac{1}{2}(A - B)$$
$$\sin A - \sin B = 2 \cos \tfrac{1}{2}(A + B) \sin \tfrac{1}{2}(A - B)$$
$$\cos A + \cos B = 2 \cos \tfrac{1}{2}(A + B) \cos \tfrac{1}{2}(A - B)$$
$$\cos A - \cos B = 2 \sin \tfrac{1}{2}(A + B) \sin \tfrac{1}{2}(B - A)$$

Product formulae

$$\sin A \sin B = \tfrac{1}{2}\{\cos(A - B) - \cos(A + B)\}$$
$$\cos A \cos B = \tfrac{1}{2}\{\cos(A - B) + \cos(A + B)\}$$
$$\sin A \cos B = \tfrac{1}{2}\{\sin(A - B) + \sin(A + B)\}$$

Powers of trigonometric functions

$$\sin^2 A = \tfrac{1}{2} - \tfrac{1}{2}\cos 2A$$
$$\cos^2 A = \tfrac{1}{2} + \tfrac{1}{2}\cos 2A$$
$$\sin^3 A = \tfrac{3}{4}\sin A - \tfrac{1}{4}\sin 3A$$
$$\cos^3 A = \tfrac{3}{4}\cos A + \tfrac{1}{4}\cos 3A$$

2.8.15 Co-ordinate geometry

Straight-line

General equation

$$ax + by + c = 0$$

m = gradient
c = intercept on the y-axis

Gradient equation

$$y = mx + c$$

Intercept equation

$$\frac{x}{A} + \frac{y}{B} = 1$$

A = intercept on the x-axis
B = intercept on the y-axis

Perpendicular equation

$$x \cos \alpha + y \sin \alpha = p$$

p = length of perpendicular from the origin to the line
α = angle that the perpendicular makes with the x-axis

The distance between two points $P(x_1, y_1)$ and $Q(x_2, y_2)$ and is given by:

$$PQ = \sqrt{(x_1 - x_2)^2 + (y_1 - y_2)^2}$$

The equation of the line joining two points (x_1, y_1) and (x_2, y_2) is given by:

$$\frac{y - y_1}{y_1 - y_2} = \frac{x - x_1}{x_1 - x_2}$$

Circle

General equation $x^2 - y^2 + 2gx + 2fy + c = 0$

The centre has co-ordinates $(-g, -f)$

The radius is $r = \sqrt{g^2 + f^2 - c}$

The equation of the tangent at (x_1, y_1) to the circle is:

$$xx_1 + yy_1 + g(x + x_1) + f(y + y_1) + c = 0$$

The length of the tangent from to the circle is:

$$t^2 = x_1^2 + y_1^2 + 2gx_1 + 2fy_1 + c$$

Parabola (see Figure 2.10)

$$\text{Eccentricity} = e = \frac{SP}{PD} = 1$$

With focus $S(a, 0)$ the equation of a parabola is $y^2 = 4ax$.

The parametric form of the equation is $x = at^2$, $y = 2at$.

The equation of the tangent at (x_1, y_1) is $yy_1 = 2a(x + x_1)$.

Ellipse (see Figure 2.11)

$$\text{Eccentricity } e = \frac{SP}{PD} < 1$$

The equation of an ellipse is $\frac{x^2}{a^2} + \frac{y^2}{b^2} = 1$ where $b^2 = a^2(1 - e^2)$.

The equation of the tangent at (x_1, y_1) is

$$\frac{xx^1}{a^2} + \frac{yy^1}{b^2} = 1.$$

The parametric form of the equation of an ellipse is $x = a\cos\theta$, $y = b\sin\theta$, where θ is the eccentric angle.

Hyperbola (see Figure 2.12)

$$\text{Eccentricity } e = \frac{SP}{PD} > 1$$

The equation of a hyperbola is $\frac{x^2}{a^2} - \frac{y^2}{b^2} = 1$ where $b^2 = a^2(e^2 - 1)$.

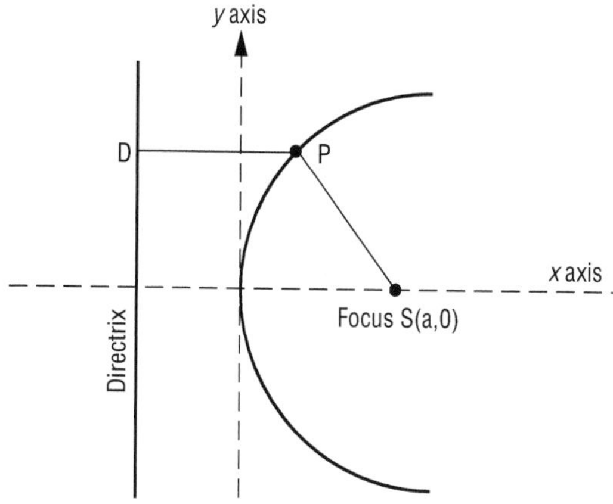

Fig. 2.10 Parabola

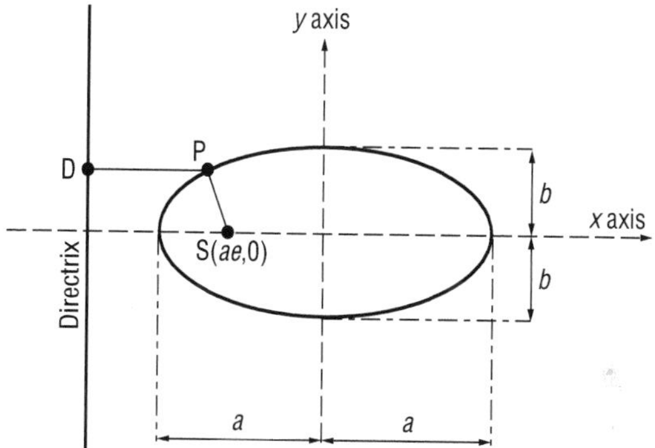

Fig. 2.11 Ellipse

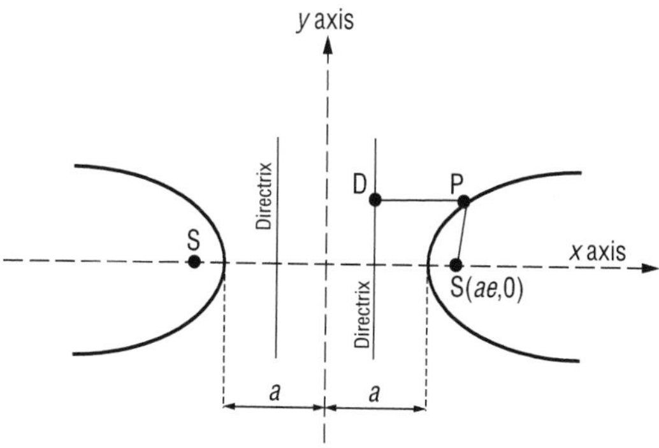

Fig. 2.12 Hyperbola

The parametric form of the equation is $x = a\sec\theta$, $y = b\tan\theta$ where θ s the eccenteric angle.

The equation of the tangent at (x_1, y_1) is

$$\frac{xx_1}{a^2} - \frac{yy_1}{b^2} = 1.$$

Sine Wave (see Figure 2.13)

$y = a\sin(bx + c)$

$y = a\cos(bx + c') = a\sin(bx + c)$ (where $c = c' + \pi/2$)

$y = m\sin bx + n\cos bx = a\sin(bx + c)$

where $a = \sqrt{m^2 + n^2}$, $c = \tan^{-1}(n/m)$.

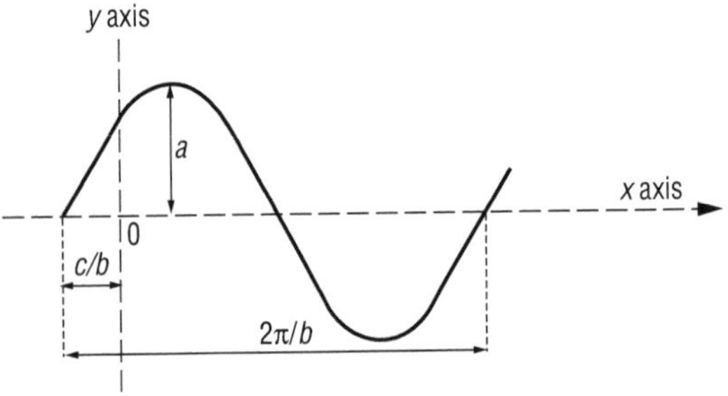

Fig. 2.13 Sine wave

Helix (see Figure 2.14)
A helix is a curve generated by a point moving on a cylinder with the distance it transverses parallel to the axis of the cylinder being proportional to the angle of rotation about the axis:

$x = a\cos\theta$
$y = a\sin\theta$
$z = k\theta$

where a = radius of cylinder, $2\pi k$ = pitch.

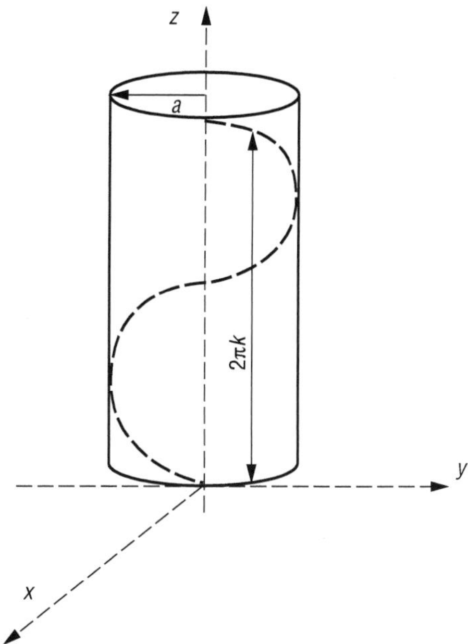

Fig. 2.14 Helix

2.9 Useful references and standards

For links to 'The Reference Desk' – a website containing over 6000 on-line units conversions 'calculators' – go to: www.flinthills.com/~ramsdale/EngZone/refer.htm

United States Metric Association, go to: http://lamar.colostate.edu/~hillger/ This site contains links to over 20 units-related sites. For guidance on correct units usage go to: http://lamar.colostate.edu/~hillger/correct.htm

Standards

1. ASTM/IEEE SI 10: 1997: *Use of the SI system of units* (replaces ASTM E380 and IEEE 268).
2. Taylor, B.N. *Guide for the use of the International System of units (SI)*: 1995. NIST special publication No 8111.

3. Federal Standard 376B: 1993: *Preferred Metric Units for general use by the Federal Government*. General Services Administration, Washington DC, 20406.

Section 3

Symbols and notations

3.1 Parameters and constants

See Table 3.1.

Table 3.1 Important parameters and constants

Planck's constant (h)	$6.6260755 \times 10^{-34}$ J s
Universal gas constant (R)	8.314510 J/mol/K
Stefan–Boltzmann constant (σ)	5.67051×10^{-8} W/m^2 K^4
Acceleration due to gravity (g)	9.80665 m/s^2 (32.17405 ft/s^2)
Absolute zero	–273.16°C (–459.688°F)
Volume of 1 kg mol of ideal gas at 1 atm, 0°C	22.41 m^3
Avagadro's number (N)	6.023×10^{26}/kg mol
Speed of sound at sea level (a_0)	340.29 m/s (1116.44 ft/sec)
Air pressure at sea level (p_0)	760 mmHg = 1.01325×10^5 N/m^2 = 2116.22 lb/ft^2
Air temperature at sea level (T_0)	15.0°C (59°F)
Air density at sea level (ρ_0)	1.22492 kg/m^3 (0.002378 slug/ft^3)
Air dynamic viscosity at sea level (μ_o)	1.4607×10^{-5} m^2/s (1.5723×10^{-4} ft^2/s)

3.2 Weights of gases

See Table 3.2.

Table 3.2 Weights of gases

Gas	*kg/m^3*	*lb/ft^3*
Air	1.22569	0.07651 (at 59.0°C)
Carbon dioxide	1.97702	0.12341
Carbon monoxide	1.25052	0.07806
Helium	0.17846	0.01114
Hydrogen	0.08988	0.005611
Nitrogen	1.25068	0.07807
Oxygen	1.42917	0.089212

All values at atmospheric pressure and 0°C.

3.3 Densities of liquids at 0°C

See Table 3.3.

Table 3.3 Densities of liquids at 0°C

Liquid		*kg/m³*	*lb/ft³*	*Specific gravity*
Water		1000	62.43	1
Sea water		1025	63.99	1.025
Jet fuel	JP 1	800	49.9	0.8
	JP 3	775	48.4	0.775
	JP 4	785	49	0.785
	JP 5	817	51	0.817
Kerosine		820	51.2	0.82
Alcohol		801	50	0.801
Gasoline (petrol)		720	44.9	0.72
Benzine		899	56.12	0.899
Oil		890	55.56	0.89

3.4 Notation: aerodynamics and fluid mechanics

See Table 3.4.

Table 3.4 Notation: aerodynamics and fluid mechanics

The complexity of aeronautics means that symbols may have several meanings, depending on the context in which they are used.

a	Lift curve slope. Acceleration or deceleration. Local speed of sound. Radius of vortex core.
a'	Inertial or absolute acceleration.
a_0	Speed of sound at sea level. Tailplane zero incidence lift coefficient.
a_1	Tailplane lift curve slope.
a_2	Elevator lift curve slope.
a_3	Elevator tab lift curve slope.
a_∞	Lift curve slope of an infinite span wing.
a_h	Local lift curve slope at spanwise co-ordinate h.
a_y	Local lift curve slope at spanwise co-ordinate y.
ac	Aerodynamic centre.
A	Aspect ratio. Moment of inertia. Area.
$\mathbf{A}$	State matrix.
AF	Activity factor of propeller.
b	Total wing-span (= $2s$). Hinge moment coefficient slope. Rotational factor in propeller theory. General width.
b_1	Elevator hinge moment derivative with respect to α_T.
b_2	Elevator hinge moment derivative with respect to η.

Table 3.4 ***Continued***

b_3	Elevator hinge moment derivative with respect to β_η.
B	Input matrix. Number of blades on a propeller.
c	Wing chord. Viscous damping coefficient. Pitot tube coefficient.
c_0	Root chord.
c_t	Tip chord.
c_y	Local chord at spanwise co-ordinate y.
cg	Centre of gravity.
cp	Centre of pressure.
C	Output matrix.
C_C	Coefficient of contraction.
C_D	Total drag coefficient.
C_{DO}	Zero lift drag coefficient.
C_f	Frictional drag coefficient.
C_L	Lift coefficient.
C_{LW}	Wing lift coefficient.
C_{LT}	Tailplane lift coefficient.
C_H	Elevator hinge moment coefficient.
C_m	Pitching moment coefficient.
C_{MO}	Pitching moment coefficient about aerodynamic centre of wing.
C_n	Yawing moment coefficient.
C_p	Pressure coefficient. Power coefficient for propellers.
C_R	Resultant force coefficient.
C_v	Coefficient of velocity.
CP	Centre of pressure.
D	Drag. Propeller diameter.
D'	Drag in a lateral-directional perturbation.
D	Direction cosine matrix. Direct matrix.
D_c	Camber drag.
D_f	Friction drag.
D_p	Pressure drag.
D_α	Incidence drag.
f	Coefficient of friction.
F	Aerodynamic force. Feed-forward path transfer function. Fractional flap chord.
F_c	Aerodynamic force due to camber.
F_r	Froude number.
F_α	Aerodynamic force due to incidence.
F_η	Elevator control force
g	Acceleration due to gravity.
G	Controlled system transfer function.
h	Height. Centre of gravity position on reference chord. Enthalpy (specific).
h_0	Aerodynamic centre position.
h_F	Fin height co-ordinate above roll axis.
h_m	Controls-fixed manoeuvre point position on reference chord.
h'_m	Controls-free manoeuvre point position on reference chord.

Table 3.4 *Continued*

h_n	Controls-fixed neutral point position on reference chord.
h'_n	Control-free neutral point position on reference chord.
H	Hinge moment. Feedback path transfer function. Total pressure. Shape factor.
H_F	Fin span measured perpendicular to the roll axis.
H_m	Controls fixed manoeuvre margin.
H''_m	Controls free manoeuvre margin.
i_x	Moment of inertia in roll (dimensionless).
i_y	Moment of inertia in pitch (dimensionless).
i_z	Moment of inertia in yaw (dimensionless).
I''	Normalized inertia.
I_x	Moment of inertia in roll.
I_y	Moment of inertia in pitch.
I_z	Moment of inertia in yaw.
J	Propeller ratio of advance. Moment of inertia.
j (or i)	The imaginary operator ($\sqrt{-1}$).
k	Spring stiffness coefficient. Lift-dependent drag factor. Interference factor.
k_{cp}	Centre of pressure coefficient.
k_d	Cavitation number.
k_q	Pitch rate transfer function gain constant.
k_u	Axial velocity transfer function gain constant.
k_w	Normal velocity transfer function gain constant.
k_θ	Pitch attitude transfer function gain constant.
k_τ	Turbo-jet engine gain constant.
K	Feedback gain. Circulation. Bulk modulus.
$\mathbf{K}$	Feedback gain matrix.
K_0	Circulation at wing mid-section.
K_n	Controls-fixed static stability margin.
K'_n	Controls-free static stability margin.
l	Lift per unit span.
l_d	Disc loading (helicopter).
l_f	Fin arm.
l_t	Tail arm.
L	Lift. Rolling moment. Temperature lapse rate.
L_c	Lift due to camber.
L_w	Wing lift.
L_F	Fin lift.
L_T	Tailplane lift.
L_α	Lift due to incidence.
m	Mass. Strength of a source or sink (fluid mechanics). Hydraulic depth.
m'	Rate of mass flow.
M	Mach number.
M_0	Free stream Mach number.
M_{crit}	Critical Mach number.
M	Pitching moment.
M_0	Wing–body pitching moment.
M_T	Tailplane pitching moment

Table 3.4 ***Continued***

n	Frequency. Number of revs per second. Polytropic exponent.
N	Yawing moment.
o	Origin of co-ordinates.
p	Roll rate perturbation. Static pressure in a fluid.
P	Power. Total pressure.
P_0	Stagnation pressure.
P_s	Static pressure.
P_t	Total pressure.
q	Pitch rate perturbation. A propeller coefficient. Discharge quantity.
Q	Dynamic pressure.
r	Yaw rate perturbation. General response variable. Radius vector.
R	Radius of turn. Resultant force. Characteristic gas constant.
Re	Reynolds number.
s	Wing semi-span. Laplace operator. Specific entropy. Distance or displacement.
S	Wing area.
S_B	Projected body side reference area.
S_F	Fin reference area.
S_T	Tailplane reference area.
t	Time. Maximum airfoil section thickness.
T	Time constant. Thrust. Temperature.
T_r	Roll time constant.
T_s	Spiral time constant.
u	Velocity component. Internal energy.
$\mathbf{u}$	Input vector.
U	Total axial velocity.
U_e	Axial component of steady equilibrium velocity.
U_E	Axial velocity component referred to datum-path earth axes.
v	Lateral velocity perturbation.
$\mathbf{v}$	Eigenvector.
V	Total lateral velocity.
V_e	Lateral component of steady equilibrium velocity.
V_E	Lateral velocity component referred to datum-path earth axes.
V_0	Steady equilibrium velocity.
V_F	Fin volume ratio.
V_R	Resultant speed.
V_S	Stalling speed.
V_T	Tailplane volume ratio.
$\mathbf{V}$	Eigenvector matrix.
w	Normal velocity perturbation. Wing loading. Downwash velocity.
W	Total nomal velocity. Weight.
W_e	Normal component of steady equilibrium velocity.
W_E	Normal velocity component referred to datum-path earth axes.

Table 3.4 ***Continued***

x	Longitudinal co-ordinate in axis system.
$\mathbf{x}$	State vector.
X	Axial force component.
y	Lateral co-ordinate.
y_B	Lateral body 'drag' coefficient.
$\mathbf{y}$	Output vector.
Y	Lateral force component.
z	Normal co-ordinate in axis system. Spanwise co-ordinate.
$\mathbf{z}$	Transformed state vector.
Z	Normal force component.
Greek symbols	
α	Angle of incidence or attack. Acceleration (angular).
α'	Incidence perturbation.
α_e	Equilibrium incidence.
α_T	Local tailplane incidence.
β	Sideslip angle perturbation. Compressibility.
β_e	Equilibrium sideslip angle.
β_η	Elevator trim tab angle.
γ	Flight path angle perturbation.
γ_e	Equilibrium flight path angle.
Γ	Wing dihedral angle (half). Circulation. Strength of vortex.
δ	Airfoil section camber. Boundary layer thickness.
δm	Mass increment.
ϵ	Throttle lever angle. Downwash angle.
ζ	Rudder angle perturbation. Damping ratio. Vorticity.
η	Efficiency.
θ	Pitch angle perturbation. Angle.
θ_e	Equilibrium pitch angle. Angular co-ordinate (polar). Propeller helix angle.
λ	Eigenvalue. Wavelength. Friction coefficient in a pipe.
Λ	Wing sweep angle.
μ	Viscosity (dynamic).
μ_1	Longitudinal relative density factor.
μ_2	Lateral relative density factor.
ν	Viscosity (kinematic).
ξ	Aileron angle perturbation.
ρ	Density.
σ	Aerodynamic time parameter. Tensile stress.
τ	Engine thrust perturbation. Shear stress.
ϕ	Phase angle. A general angle.
Φ	State transition matrix.
Ψ	Yaw angle perturbation. Stream function.
ω	Natural frequency. Angular velocity.
ω_b	Bandwidth frequency.
ω_n	Damped natural frequency.

Table 3.4 ***Continued***

Subscripts	
0	Datum axes. Normal earth-fixed axes. Straight/level flight. Free stream flow conditions. Sea level.
1/4	Quarter chord.
2	Double or twice.
∞	Infinity condition.
a	Aerodynamic. Available.
b	Aeroplane body axes. Bandwidth.
c	Chord. Compressible flow. Camber line.
D	Drag.
e	Equilibrium.
E	Earth axes.
F	Fin.
g	Gravitational. Ground.
h	Horizontal.
H	Elevator hinge moment.
i	Incompressible. Ideal.
l	Rolling moment.
LE	Leading edge.
L	Lift.
m	Pitching moment. Manoeuvre.
n	Damped natural frequency.
n	Neutral point. Yawing moment.
p	Power. Phugoid.
p	Roll rate.
q	Pitch rate.
r	Roll mode.
r	Yaw rate.
s	Short period pitching oscillation. Spiral. Stagnation. Surface.
t	Tangential.
TE	Trailing edge.
T	Tailplane.
u	Axial velocity.
U	Upper.
v	Lateral velocity.
V	Vertical.
w	Wing.
w	Normal velocity.
x	ox axis.
y	oy axis.
z	oz axis.
α	Angle of attack or incidence.
ϵ	Throttle lever.
ζ	Rudder.
η	Elevator.
θ	Pitch.
ξ	Ailerons.
τ	Thrust.

3.5 The International Standard Atmosphere (ISA)

The ISA is an internationally agreed set of assumptions for conditions at mean sea level and the variations of atmosphere conditions with altitude. In the troposphere (up to 11 000 m), temperature varies with altitude at a standard lapse rate L, measured in K (or °C) per metre. Above 11 000 m, it is assumed that temperature does not vary with height (Figure 3.1).

So, in the troposphere:

Temperature variation is given by:

$$T = T_0 - Lh$$

Pressure is given by: $\dfrac{p_2}{p_1} = \left(\dfrac{T_2}{T_1}\right)^{5.256}$

where T = temperature at an altitude h (m)
T_0 = absolute temperature at mean sea level (K)
L = lapse rate in K/m
p = pressure at an altitude

The lapse rate L in the ISA is 6.5 K/km.

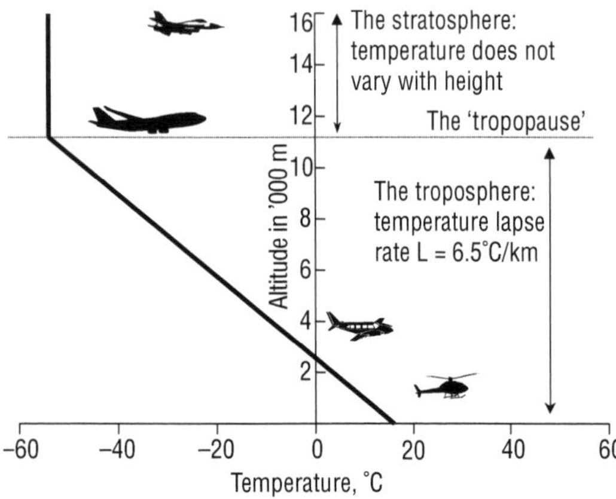

Fig. 3.1 The ISA; variation of temperature with altitude

In the stratosphere $T = T_S$ = constant so:

$$\frac{p_1}{p_2} = \frac{\rho_1}{\rho_2} \text{ and } \frac{p}{\rho} = RT$$

where R is the universal gas constant: R = 287.26 J/kg K

Table 3.5 shows the international standard atmosphere (ISA). Table 3.6 shows the lesser used US (COESA) standard atmosphere.

Table 3.5 International standard atmosphere (sea level conditions)

Property	*Metric value*	*Imperial value*
Pressure (p)	101 304 Pa	2116.2 lbf/ft^2
Density (ρ)	1.225 kg/m^3	0.002378 slug/ft^3
Temperature (t)	15°C or 288.2 K	59°F or 518.69°R
Speed of sound (a)	340 m/s	1116.4 ft/s
Viscosity (μ)	1.789×10^{-5} kg/m s	3.737×10^{-7} slug/ft s
Kinematic viscosity (ν)	1.460×10^{-5} m^2/s	1.5723×10^{-4} ft^2/s
Thermal conductivity	0.0253 J/m s/K	0.01462 BTU/ft h°F
Gas constant (R)	287.1 J/kg K	1715.7 ft lb/slug/°R
Specific heat (C_p)	1005 J/kg K	6005 ft lb/slug/°R
Specific heat (C_v)	717.98 J/kg K	4289 ft lb/slug/°R
Ratio of specific heats (γ)	1.40	1.40
Gravitational acceleration (g)	9.80665 m/s^2	32.174 ft/s^2

Table 3.5 *Continued*

Altitude		*Temperature*	*Pressure ratio*	*Density ratio*	*Dynamic viscosity ratio*	*Kinematic viscosity ratio*	*a*
(m)	*(ft)*	*(°C)*	(p/p_o)	(ρ/ρ_o)	(μ/μ_o)	(μ/μ_o)	*(m/s)*
0	0	15.2	1.0000	1.0000	1.0000	1.0000	340.3
152	500	14.2	0.9821	0.9855	0.9973	1.0121	339.7
304	1000	13.2	0.9644	0.9711	0.9947	1.0243	339.1
457	1500	12.2	0.9470	0.9568	0.9920	1.0367	338.5
609	2000	11.2	0.9298	0.9428	0.9893	1.0493	338.0
762	2500	10.2	0.9129	0.9289	0.9866	1.0622	337.4
914	3000	9.3	0.8962	0.9151	0.9839	1.0752	336.8
1066	3500	8.3	0.8798	0.9015	0.9812	1.0884	336.2
1219	4000	7.3	0.8637	0.8881	0.9785	1.1018	335.6
1371	4500	6.3	0.8477	0.8748	0.9758	1.1155	335.0
1524	5000	5.3	0.8320	0.8617	0.9731	1.1293	334.4
1676	5500	4.3	0.8166	0.8487	0.9704	1.1434	333.8
1828	6000	3.3	0.8014	0.8359	0.9677	1.1577	333.2
1981	6500	2.3	0.7864	0.8232	0.9649	1.1722	332.6
2133	7000	1.3	0.7716	0.8106	0.9622	1.1870	332.0
2286	7500	0.3	0.7571	0.7983	0.9595	1.2020	331.4

2438	8000	–0.6	0.7428	0.7860	0.9567	1.2172	330.8
2590	8500	–1.6	0.7287	0.7739	0.9540	1.2327	330.2
2743	9000	–2.6	0.7148	0.7620	0.9512	1.2484	329.6
2895	9500	–3.6	0.7012	0.7501	0.9485	1.2644	329.0
3048	10000	–4.6	0.6877	0.7385	0.9457	1.2807	328.4
3200	10500	–5.6	0.6745	0.7269	0.9430	1.2972	327.8
3352	11000	–6.6	0.6614	0.7155	0.9402	1.3140	327.2
3505	11500	–7.6	0.6486	0.7043	0.9374	1.3310	326.6
3657	12000	–8.6	0.6360	0.6932	0.9347	1.3484	326.0
3810	12500	–9.6	0.6236	0.6822	0.9319	1.3660	325.4
3962	13000	–10.6	0.6113	0.6713	0.9291	1.3840	324.7
4114	13500	–11.5	0.5993	0.6606	0.9263	1.4022	324.1
4267	14000	–12.5	0.5875	0.6500	0.9235	1.4207	323.5
4419	14500	–13.5	0.5758	0.6396	0.9207	1.4396	322.9
4572	15000	–14.5	0.5643	0.6292	0.9179	1.4588	322.3
4724	15500	–15.5	0.5531	0.6190	0.9151	1.4783	321.7
4876	16000	–16.5	0.5420	0.6090	0.9123	1.4981	321.0
5029	16500	–17.5	0.5311	0.5990	0.9094	1.5183	320.4
5181	17000	–18.5	0.5203	0.5892	0.9066	1.5388	319.8
5334	17500	–19.5	0.5098	0.5795	0.9038	1.5596	319.2
5486	18000	–20.5	0.4994	0.5699	0.9009	1.5809	318.5

Table 3.5 ***Continued***

Altitude		*Temperature*	*Pressure ratio*	*Density ratio*	*Dynamic viscosity ratio*	*Kinematic viscosity ratio*	*a*
(m)	*(ft)*	*(°C)*	(p/p_o)	(ρ/ρ_o)	(μ/μ_o)	(μ/μ_o)	*(m/s)*
5638	18500	–21.5	0.4892	0.5604	0.8981	1.6025	317.9
5791	19000	–22.4	0.4791	0.5511	0.8953	1.6244	317.3
5943	19500	–23.4	0.4693	0.5419	0.8924	1.6468	316.7
6096	20000	–24.4	0.4595	0.5328	0.8895	1.6696	316.0
6248	20500	–25.4	0.4500	0.5238	0.8867	1.6927	315.4
6400	21000	–26.4	0.4406	0.5150	0.8838	1.7163	314.8
6553	21500	–27.4	0.4314	0.5062	0.8809	1.7403	314.1
6705	22000	–28.4	0.4223	0.4976	0.8781	1.7647	313.5
6858	22500	–29.4	0.4134	0.4891	0.8752	1.7895	312.9
7010	23000	–30.4	0.4046	0.4806	0.8723	1.8148	312.2
7162	23500	–31.4	0.3960	0.4723	0.8694	1.8406	311.6
7315	24000	–32.3	0.3876	0.4642	0.8665	1.8668	311.0
7467	24500	–33.3	0.3793	0.4561	0.8636	1.8935	310.3
7620	25000	–34.3	0.3711	0.4481	0.8607	1.9207	309.7
7772	25500	–35.3	0.3631	0.4402	0.8578	1.9484	309.0

7924	26000	–36.3	0.3552	0.4325	0.8548	1.9766	308.4
8077	26500	–37.3	0.3474	0.4248	0.8519	2.0053	307.7
8229	27000	–38.3	0.3398	0.4173	0.8490	2.0345	307.1
8382	27500	–39.3	0.3324	0.4098	0.8460	2.0643	306.4
8534	28000	–40.3	0.3250	0.4025	0.8431	2.0947	305.8
8686	28500	–41.3	0.3178	0.3953	0.8402	2.1256	305.1
8839	29000	–42.3	0.3107	0.3881	0.8372	2.1571	304.5
8991	29500	–43.2	0.3038	0.3811	0.8342	2.1892	303.8
9144	30000	–44.2	0.2970	0.3741	0.8313	2.2219	303.2
9296	30500	–45.2	0.2903	0.3673	0.8283	2.2553	302.5
9448	31000	–46.2	0.2837	0.3605	0.8253	2.2892	301.9
9601	31500	–47.2	0.2772	0.3539	0.8223	2.3239	301.2
9753	32000	–48.2	0.2709	0.3473	0.8194	2.3592	300.5
9906	32500	–49.2	0.2647	0.3408	0.8164	2.3952	299.9
10058	33000	–50.2	0.2586	0.3345	0.8134	2.4318	299.2
10210	33500	–51.2	0.2526	0.3282	0.8104	2.4692	298.6
10363	34000	–52.2	0.2467	0.3220	0.8073	2.5074	297.9
10515	34500	–53.2	0.2410	0.3159	0.8043	2.5463	297.2
10668	35000	–54.1	0.2353	0.3099	0.8013	2.5859	296.5
10820	35500	–55.1	0.2298	0.3039	0.7983	2.6264	295.9
10972	36000	–56.1	0.2243	0.2981	0.7952	2.6677	295.2

Table 3.5 *Continued*

Altitude		*Temperature*	*Pressure ratio*	*Density ratio*	*Dynamic viscosity ratio*	*Kinematic viscosity ratio*	*a*
(m)	*(ft)*	*(°C)*	(p/p_o)	(ρ/ρ_o)	(μ/μ_o)	(μ/μ_o)	*(m/s)*
10999	36089	–56.3	0.2234	0.2971	0.7947	2.6751	295.1
11277	37000	–56.3	0.2138	0.2843	0.7947	2.7948	295.1
11582	38000	–56.3	0.2038	0.2710	0.7947	2.9324	295.1
11887	39000	–56.3	0.1942	0.2583	0.7947	3.0768	295.1
12192	40000	–56.3	0.1851	0.2462	0.7947	3.2283	295.1
12496	41000	–56.3	0.1764	0.2346	0.7947	3.3872	295.1
12801	42000	–56.3	0.1681	0.2236	0.7947	3.5540	295.1
13106	43000	–56.3	0.1602	0.2131	0.7947	3.7290	295.1
13411	44000	–56.3	0.1527	0.2031	0.7947	3.9126	295.1
13716	45000	–56.3	0.1456	0.1936	0.7947	4.1052	295.1
14020	46000	–56.3	0.1387	0.1845	0.7947	4.3073	295.1
14325	47000	–56.3	0.1322	0.1758	0.7947	4.5194	295.1
14630	48000	–56.3	0.1260	0.1676	0.7947	4.7419	295.1
14935	49000	–56.3	0.1201	0.1597	0.7947	4.9754	295.1
15240	50000	–56.3	0.1145	0.1522	0.7947	5.2203	295.1

15544	51000	–56.3	0.1091	0.1451	0.7947	5.4773	295.1
15849	52000	–56.3	0.1040	0.1383	0.7947	5.7470	295.1
16154	53000	–56.3	0.9909^{-1}	0.1318	0.7947	6.0300	295.1
16459	54000	–56.3	0.9444^{-1}	0.1256	0.7947	6.3268	295.1
16764	55000	–56.3	0.9001^{-1}	0.1197	0.7947	6.6383	295.1
17068	56000	–56.3	0.8579^{-1}	0.1141	0.7947	6.9652	295.1
17373	57000	–56.3	0.8176^{-1}	0.1087	0.7947	7.3081	295.1
17678	58000	–56.3	0.7793^{-1}	0.1036	0.7947	7.6679	295.1
17983	59000	–56.3	0.7427^{-1}	0.9878^{-1}	0.7947	8.0454	295.1
18288	60000	–56.3	0.7079^{-1}	0.9414^{-1}	0.7947	8.4416	295.1
18592	61000	–56.3	0.6746^{-1}	0.8972^{-1}	0.7947	8.8572	295.1
18897	62000	–56.3	0.6430^{-1}	0.8551^{-1}	0.7947	9.2932	295.1
19202	63000	–56.3	0.6128^{-1}	0.8150^{-1}	0.7947	9.7508	295.1
19507	64000	–56.3	0.5841^{-1}	0.7768^{-1}	0.7947	10.231	295.1
19812	65000	–56.3	0.5566^{-1}	0.7403^{-1}	0.7947	10.735	295.1
20116	66000	–56.3	0.5305^{-1}	0.7056^{-1}	0.7947	11.263	295.1
20421	67000	–56.3	0.5056^{-1}	0.6725^{-1}	0.7947	11.818	295.1
20726	68000	–56.3	0.4819^{-1}	0.6409^{-1}	0.7947	12.399	295.1
21031	69000	–56.3	0.4593^{-1}	0.6108^{-1}	0.7947	13.010	295.1
21336	70000	–56.3	0.4377^{-1}	0.5822^{-1}	0.7947	13.650	295.1

Table 3.6 US/COESA atmosphere (SI units)

Alt (km)	$\rho/\rho o$	p/p_o	t/t_o	*temp. (K)*	*press. (N/m²)*	*dens. (kg/m³)*	*a (m/s)*	μ *(10⁻⁶ kg/ms)*	ν *(m²/s)*
–2	1.2067E+0	1.2611E+0	1.0451	301.2	1.278E+5	1.478E+0	347.9	18.51	1.25E–5
0	1.0000E+0	1.0000E+0	1.0000	288.1	1.013E+5	1.225E+0	340.3	17.89	1.46E–5
2	8.2168E–1	7.8462E–1	0.9549	275.2	7.950E+4	1.007E+0	332.5	17.26	1.71E–5
4	6.6885E–1	6.0854E–1	0.9098	262.2	6.166E+4	8.193E–1	324.6	16.61	2.03E–5
6	5.3887E–1	4.6600E–1	0.8648	249.2	4.722E+4	6.601E–1	316.5	15.95	2.42E–5
8	4.2921E–1	3.5185E–1	0.8198	236.2	3.565E+4	5.258E–1	308.1	15.27	2.90E–5
10	3.3756E–1	2.6153E–1	0.7748	223.3	2.650E+4	4.135E–1	299.5	14.58	3.53E–5
12	2.5464E–1	1.9146E–1	0.7519	216.6	1.940E+4	3.119E–1	295.1	14.22	4.56E–5
14	1.8600E–1	1.3985E–1	0.7519	216.6	1.417E+4	2.279E–1	295.1	14.22	6.24E–5
16	1.3589E–1	1.0217E–1	0.7519	216.6	1.035E+4	1.665E–1	295.1	14.22	8.54E–5
18	9.9302E–2	7.4662E–2	0.7519	216.6	7.565E+3	1.216E–1	295.1	14.22	1.17E–4
20	7.2578E–2	5.4569E–2	0.7519	216.6	5.529E+3	8.891E–2	295.1	14.22	1.60E–4
22	5.2660E–2	3.9945E–2	0.7585	218.6	4.047E+3	6.451E–2	296.4	14.32	2.22E–4
24	3.8316E–2	2.9328E–2	0.7654	220.6	2.972E+3	4.694E–2	297.7	14.43	3.07E–4
26	2.7964E–2	2.1597E–2	0.7723	222.5	2.188E+3	3.426E–2	299.1	14.54	4.24E–4
28	2.0470E–2	1.5950E–2	0.7792	224.5	1.616E+3	2.508E–2	300.4	14.65	5.84E–4
30	1.5028E–2	1.1813E–2	0.7861	226.5	1.197E+3	1.841E–2	301.7	14.75	8.01E–4
32	1.1065E–2	8.7740E–3	0.7930	228.5	8.890E+2	1.355E–2	303.0	14.86	1.10E–3
34	8.0709E–3	6.5470E–3	0.8112	233.7	6.634E+2	9.887E–3	306.5	15.14	1.53E–3
36	5.9245E–3	4.9198E–3	0.8304	239.3	4.985E+2	7.257E–3	310.1	15.43	2.13E–3

38	4.3806E–3	3.7218E–3	0.8496	244.8	3.771E+2	5.366E–3	313.7	15.72	2.93E–3
40	3.2615E–3	2.8337E–3	0.8688	250.4	2.871E+2	3.995E–3	317.2	16.01	4.01E–3
42	2.4445E–3	2.1708E–3	0.8880	255.9	2.200E+2	2.995E–3	320.7	16.29	5.44E–3
44	1.8438E–3	1.6727E–3	0.9072	261.4	1.695E+2	2.259E–3	324.1	16.57	7.34E–3
46	1.3992E–3	1.2961E–3	0.9263	266.9	1.313E+2	1.714E–3	327.5	16.85	9.83E–3
48	1.0748E–3	1.0095E–3	0.9393	270.6	1.023E+2	1.317E–3	329.8	17.04	1.29E–2
50	8.3819E–4	7.8728E–4	0.9393	270.6	7.977E+1	1.027E–3	329.8	17.04	1.66E–2
52	6.5759E–4	6.1395E–4	0.9336	269.0	6.221E+1	8.055E–4	328.8	16.96	2.10E–2
54	5.2158E–4	4.7700E–4	0.9145	263.5	4.833E+1	6.389E–4	325.4	16.68	2.61E–2
56	4.1175E–4	3.6869E–4	0.8954	258.0	3.736E+1	5.044E–4	322.0	16.40	3.25E–2
58	3.2344E–4	2.8344E–4	0.8763	252.5	2.872E+1	3.962E–4	318.6	16.12	4.07E–2
60	2.5276E–4	2.1668E–4	0.8573	247.0	2.196E+1	3.096E–4	315.1	15.84	5.11E–2
62	1.9647E–4	1.6468E–4	0.8382	241.5	1.669E+1	2.407E–4	311.5	15.55	6.46E–2
64	1.5185E–4	1.2439E–4	0.8191	236.0	1.260E+1	1.860E–4	308.0	15.26	8.20E–2
66	1.1668E–4	9.3354E–5	0.8001	230.5	9.459E+0	1.429E–4	304.4	14.97	1.05E–1
68	8.9101E–5	6.9593E–5	0.7811	225.1	7.051E+0	1.091E–4	300.7	14.67	1.34E–1
70	6.7601E–5	5.1515E–5	0.7620	219.6	5.220E+0	8.281E–5	297.1	14.38	1.74E–1
72	5.0905E–5	3.7852E–5	0.7436	214.3	3.835E+0	6.236E–5	293.4	14.08	2.26E–1
74	3.7856E–5	2.7635E–5	0.7300	210.3	2.800E+0	4.637E–5	290.7	13.87	2.99E–1
76	2.8001E–5	2.0061E–5	0.7164	206.4	2.033E+0	3.430E–5	288.0	13.65	3.98E–1
78	2.0597E–5	1.4477E–5	0.7029	202.5	1.467E+0	2.523E–5	285.3	13.43	5.32E–1
80	1.5063E–5	1.0384E–5	0.6893	198.6	1.052E+0	1.845E–5	282.5	13.21	7.16E–1
82	1.0950E–5	7.4002E–6	0.6758	194.7	7.498E–1	1.341E–5	279.7	12.98	9.68E–1
84	7.9106E–6	5.2391E–6	0.6623	190.8	5.308E–1	9.690E–6	276.9	12.76	1.32E+0
86	5.6777E–6	3.6835E–6	0.6488	186.9	3.732E–1	6.955E–6	274.1	12.53	1.80E+0

Section 4

Aeronautical definitions

4.1 Forces and moments

Forces and moments play an important part in the science of aeronautics. The basic definitions are:

Weight force (W)
: Weight of aircraft acting vertically downwards.

Aerodynamic force
: Force exerted (on an aircraft) by virtue of the diversion of an airstream from its original path. It is divided into three components: lift, drag and lateral.

Lift force (L)
: Force component perpendicularly 'upwards' to the flight direction.

Drag force (D)
: Force component in the opposite direction to flight. Total drag is subdivided into *pressure* drag and *surface friction* drag.

Pressure drag
: Force arising from resolved components of normal pressure. Pressure drag is subdivided into boundary layer pressure or *form* drag, vortex or *induced* drag, and *wave* drag.

Surface friction drag
: Force arising from surface or skin friction between a surface and a fluid.

Pitching moment (M)
: Moment tending to raise the nose of an aircraft up or down. It acts in the plane defined by the lift force and drag force.

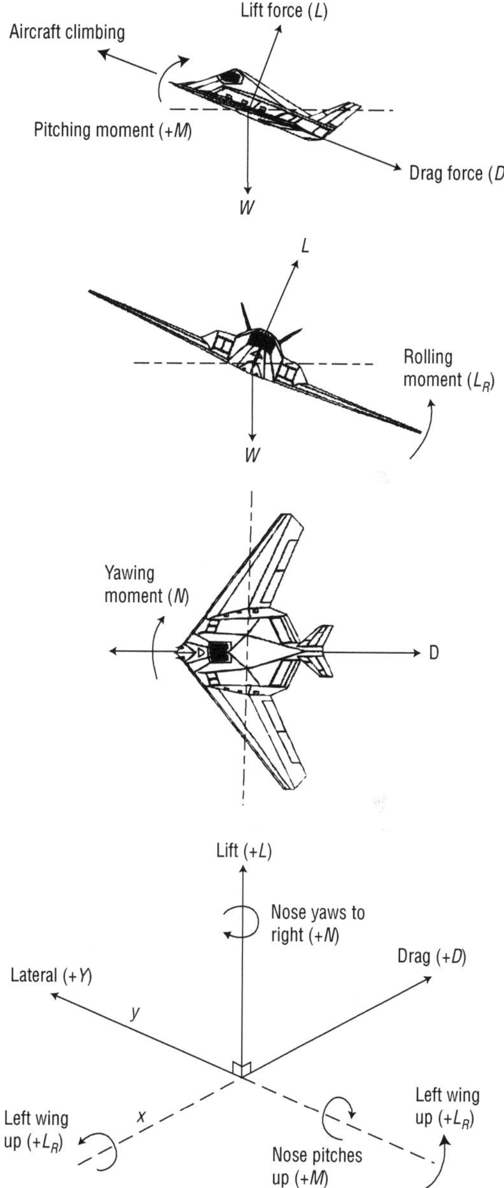

Fig. 4.1 Forces, moments and motions

Rolling moment (L_R)
Moment tending to roll an aircraft about its nose-to-tail axis (i.e. to raise or lower the wing tips).

Yawing moment (N)
Moment tending to swing the nose of an aircraft to the left or right of its direction of flight.

Figure 4.1 shows the basic sign conventions that are used. Motions are often also referred to by their relation to x-, y-, z-axes: See Table 4.1.

Table 4.1 The general axis system

Axis	*Moment*	*Moment of inertia*	*Angular displacement*
x	L_R (roll)	I_x	ϕ
y	M (pitch)	I_y	θ
z	N (yaw)	I_z	ψ

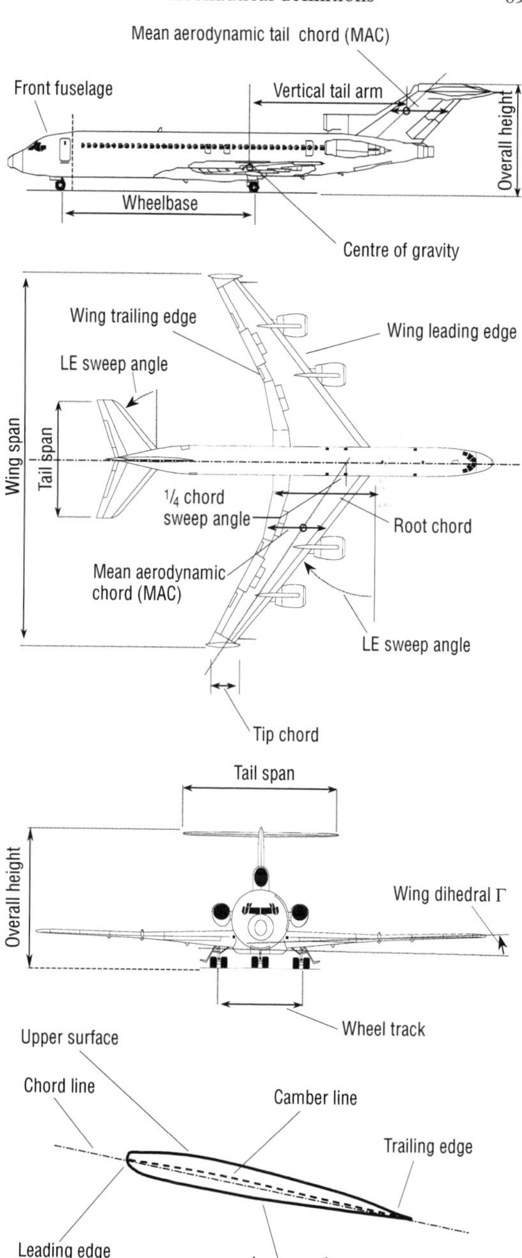

Fig. 4.2 Basic aircraft terminology

4.2 Basic aircraft terminology

Table 4.2 Basic aircraft terminology (see also Figure 4.2)

Term	Definition
Aspect ratio (A)	A measurement of the 'narrowness' of the wing form.
Camber line	A line joining the locus of points situated midway between the upper and lower surfaces of a wing.
Dihedral (2Γ)	Upward or downward (anhedral) angle of the wing.
Leading edge (LE)	Front edge of the wing.
Mean aerodynamic chord (MAC) $(\bar{c})_A$	A chord parameter defined as: $\bar{c}_A = \dfrac{\int_{-s}^{+s} c^2\,dy}{\int_{-s}^{+s} c\,dy}$
Root chord (c_O)	Chord length of the wing where it meets the fuselage.
Standard mean chord (SMC) or Geometric mean chord ($\bar{c}$)	A chord parameter given defined as $\bar{c} = S_G/b$ or S_N/b $= \dfrac{\int_{-s}^{+s} c\,dy}{\int_{-s}^{+s} dy}$
Sweepback (Λ or ϕ)	Lateral orientation of a wing measured between the lateral (y) axis and the wing leading edge Λ_{LE} or ϕ_{LE}), or the 1/4 chord position ($\Lambda_{1/4}$ or $\phi_{1/4}$), or the wing trailing edge (Λ_{TE} or ϕ_{TE}).
Tip chord (c_t)	Chord length of the wing at its tip.
Trailing edge (TE)	Rear edge of the wing.
Wing (gross) area (S_G)	The plan area of the wing, inclusive of the continuation within the fuselage.
Wing (net) area (S_N)	The plan area of the wing excluding any continuation within the fuselage.
Wing plan form	The shape of the plan view of the wing.
Wingspan (b)	Distance between the extreme tips of the wings.

4.3 Helicopter terminology

Table 4.3 Helicopter terminology and acronyms

AAH Advanced attack helicopter.
ABC Advancing-blade concept.
ACT Active-control(s) technology.
AH Attack helicopter.
ALH Advanced light helicopter.
ARTI Advanced rotorcraft technology integration.
ASW Anti-submarine warfare.

CH Cargo helicopter.
collective The mode of control in which the pitch of all rotor blades changes simultaneously (applies to main or tail rotor).
coning angle Angle between the longitudinal axis of a main-rotor blade and the tip-path plane.
cyclic The mode of control which varies blade pitch (main rotor only).
drag hinge Hinge permitting a rotor blade to pivot to the front and rear in its plane of rotation.
elastomeric bearing A bearing containing an elastomeric material (e.g. rubber).

FADEC Full-authority digital engine control.
FBL Fly-by-light; the use of optical fibres to carry coded light signals to convey main flight-control demands.
FBW Fly-by-wire; the use of electric cables to convey flight-control demands in the form of variable electric currents.
Fenestron Aérospatiale tail rotor with multiple small blades shrouded in the centre of the tail fin. Often known as 'fan in tail'.
flapping hinge Hinge which allows the tip of a rotor blade to pivot normal to the plane of rotation.

ground effect The effect of having a solid flat surface close beneath a hovering helicopter.
gyrostabilized Mounted on gimbals (pivots) and held in a constant attitude, irrespective of how the helicopter manoeuvres.

HAR Helicopter, air rescue (also ASR; Air Sea Rescue).
HELRAS Helicopter long-range active sonar.
HH Search and rescue helicopter (US).
HIGE Helicopter in ground effect.
HISOS Helicopter integrated sonics system.
HLH Heavy-lift helicopter.
hub The centre of a main or tail rotor to which the blades are attached.
HUD Head-up display; cockpit instrument which projects on to a glass screen.
IGE In ground effect; as if the helicopter had the ground immediately beneath it.

Table 4.3 ***Continued***

IMS Integrated multiplex system.
INS Inertial navigation system.
IRCM Infrared countermeasure.

lead/lag damper Cushioning buffer designed to minimize ground resonance.
LHX Light experimental helicopter programme.
LIVE Liquid inertial vibration eliminator.
LOH Light observation helicopter.

MTR Main and tail rotor.

NFOV Narrrow field of view.
nodamadic Patented form of vibration-damping system.
NOE Nap of the Earth, i.e. at the lowest safe level.
NOTAR No tail rotor.

OEI One engine inoperative.
OGE Out of ground effect.

RAST Recovery assist, securing and traversing — a system to help helicopters land on a ship's deck.
rigid rotor Rotor with a particular structure near the hub so that rotor flex replaces the function of mechanical hinges.
ROC Required operational capability.
RSRA Rotor systems research aircraft.

SCAS Stability and control augmentation system.
SH Anti-submarine helicopter (US).
sidestick Small control column at the side of the cockpit.
Starflex Trade name of advanced hingeless rotor system (Aérospatiale).
stopped-rotor aircraft A helicopter whose rotor can be slowed down and stopped in flight, its blades then behaving like four wings.
swashplate A disc either fixed or rotating on the main rotor drive shaft, which is tilted in various directions.

tip path The path in space traced out by tips of rotor blades.

UTS Universal turret system.

4.4 Common aviation terms

Table 4.4 Aviation acronyms

3/LMB	3 Light Marker Beacon
360CH	360 Channel Radio
720CH	720 Channel Radio
AC or AIR	Air Conditioning

Table 4.4 ***Continued***

ACARS	Aircraft Communication Addressing and Reporting System
AD	Airworthiness Directive
ADF	Automatic Direction Finder
AFIS	Airborne Flight Info System
AFTT	Air Frame Total Time (in hours)
AP	Autopilot
APU	Auxiliary Power Unit
ASI	Air Speed Indicator
ATIS	Automatic Terminal Information Service (a continuous broadcast of recorded non-control information in selected high activity terminal areas)
AWOS	Automatic Weather Observation Service
C of A	Certificate of Airworthiness
C/R	Counter Rotation (propellers)
CAS	Calibrated Air Speed
CHT	Cylinder Head Temperature Gauge
COM	Com Radio
CONV/MOD	Conversion/Modification (to aircraft)
DG	Directional Gyro
DME	Distance Measuring Equipment
EFIS	Electronic Flight Instrument System
EGT	Exhaust Gas Temperature Gauge
ELT	Emergency Locator Transmitter
ENC	Air Traffic Control Encoder
F/D	Flight Director
FADEC	Full Authority Digital Engine Control
FBO	Fixed Base Operation
FMS	Flight Management System
G/S	Glideslope
G/W	Gross Weight
GPS	Global Positioning System
GPWS	Ground Proximity Warning System
GS	Ground Speed
HF	High Frequency Radio
HSI	Horizontal Situation Indicator
HUD	Head Up Display
IAS	Indicated Air Speed
ICE	Has Anti-Icing Equipment
IFR	Instrument Flight Rules
ILS	Instrument Landing System
KCAS	Calibrated air speed (Knots)
KIAS	Indicated air speed (Knots)
KNOWN ICE	Certified to fly in known icing conditions
LOC	Localizer
LRF	Long Range Fuel
LRN	Loran
MLS	Microwave Landing System
N/C	Navigation and Communication Radios
NAV	Nav Radio

Table 4.4 ***Continued***

NAV/COM	Navigation and Communication Radios
NDH	No Damage History
NOTAM	Notice to Airmen (radio term)
O/H	Overhaul
OAT	Outside Air Temperature
OC	On Condition
OMEGA	VLF (Very Low Frequency) Navigation
PANTS	Fixed Gear Wheel Covers
PTT	Push to Talk
RALT	Radar Altimeter
RDR	Radar
RMI	Radio Magnetic Indicator
RNAV	Area Navigation (usually includes DME)
RSTOL	Roberson STOL Kit
SB	Service Bulletin
SFRM	(Time) Since Factory Remanufactured Overhaul
SHS	Since Hot Section
SLC	Slaved Compass
SMOH	Since Major Overhaul
SPOH	Since Propeller Overhaul
STOH	Since Top Overhaul
STOL	Short Takeoff and Landing Equipment
STORM	Stormscope
T/O	Takeoff (weight)
TAS	True Air Speed
TBO	Time Between Overhauls
TCAD	Traffic/Collision Avoidance Device
TCAS	Traffic Alert and Collision Avoidance System
TREV	Thrust Reversers
TT	Total Time
TTSN	Time Since New
TWEB	Transcribed Weather Broadcast
TXP	Transponder
Va	Safe operating speed
Vfe	Safe operating speed (flaps extended)
VFR	Visual Flight Rules
Vle	Safe operating speed (landing gear extended)
VNAV	Vertical Navigation computer
Vne	'Never exceed' speed
Vno	Maximum cruising 'normal operation' speed
VOR	Very High Frequency Omnidirectional Rangefinder
Vs	Stalling speed
VSI	Vertical Speed Indicator
Vso	Stalling speed in landing configuration
Vx	Speed for best angle of climb
Vy	Speed for best rate of climb
XPDR	Transponder

4.5 Airspace terms

The following abbreviations are in use to describe various categories of airspace.

Table 4.5 Airspace acronyms

AAL	Above airfield level
AGL	Above ground level
AIAA	Area of intense air activity
AMSL	Above mean sea level
CTA	Control area
CTZ	Control zone
FIR	Flight information region
FL	Flight level
LFA	Local flying area
MATZ	Military airfield traffic zone (UK)
MEDA	Military engineering division airfield (UK)
Min DH	Minimum descent height
SRA	Special rules airspace (area)
SRZ	Special rules zone
TMA	Terminal control area

Section 5

Basic fluid mechanics

5.1 Basic poperties

5.1.1 Basic relationships

Fluids are divided into liquids, which are virtually incompressible, and gases, which are compressible. A fluid consists of a collection of molecules in constant motion; a liquid adopts the shape of a vessel containing it whilst a gas expands to fill any container in which it is placed. Some basic fluid relationships are given in Table 5.1.

Table 5.1 Basic fluid relationships

Density (ρ)	Mass per unit volume. Units kg/m^3 (lb/in^3)
Specific gravity (s)	Ratio of density to that of water, i.e. $s = \rho/\rho_{water}$
Specific volume (v)	Reciprocal of density, i.e. s = $1/\rho$. Units m^3/kg (in^3/lb)
Dynamic viscosity (μ)	A force per unit area or shear stress of a fluid. Units Ns/m^2 (lbf.s/ft^2)
Kinematic viscosity (ν)	A ratio of dynamic viscosity to density, i.e. $\nu = \mu/\rho$. Units m^2/s (ft^2/sec)

5.1.2 Perfect gas

A perfect (or 'ideal') gas is one which follows Boyle's/Charles' law $pv = RT$ where:

p = pressure of the gas
v = specific volume
T = absolute temperature
R = the universal gas constant

Although no actual gases follow this law totally, the behaviour of most gases at temperatures well above their liquefication temperature will approximate to it and so they can be *considered* as a perfect gas.

5.1.3 Changes of state

When a perfect gas changes state its behaviour approximates to:

$$pv^n = \text{constant}$$

where n is known as the polytropic exponent.

Figure 5.1 shows the four main changes of state relevant to aeronautics: isothermal, adiabatic: polytropic and isobaric.

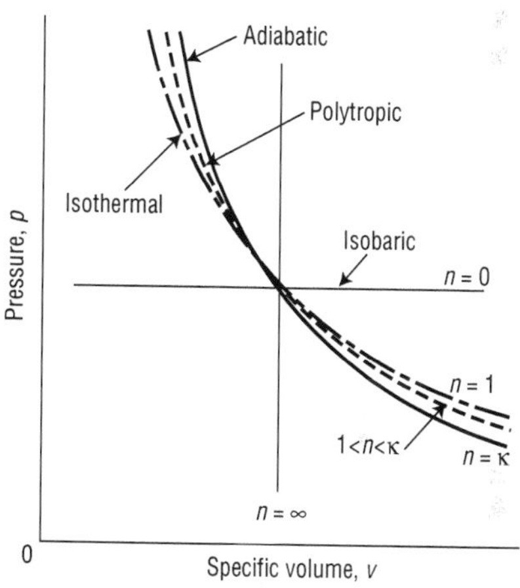

Fig. 5.1 Changes of state of a perfect gas

5.1.4 Compressibility

The extent to which a fluid can be compressed in volume is expressed using the compressibility coefficient β.

$$\beta = \frac{\Delta v/v}{\Delta p} = \frac{1}{K}$$

where Δv = change in volume
v = initial volume
Δp = change in pressure
K = bulk modulus

Also:

$$K = \rho \frac{\Delta p}{\Delta \rho} = \rho \frac{dp}{d\rho}$$

and

$$a = \sqrt{\frac{dp}{d\rho}} = \sqrt{\frac{K}{\rho}}$$

where a = the velocity of propagation of a pressure wave in the fluid

5.1.5 Fluid statics

Fluid statics is the study of fluids which are at rest (i.e. not flowing) relative to the vessel containing it. Pressure has four important characteristics:

- Pressure applied to a fluid in a closed vessel (such as a hydraulic ram) is transmitted to all parts of the closed vessel at the same value (Pascal's law).
- The magnitude of pressure force acting at any point in a static fluid is the same, irrespective of direction.
- Pressure force always acts perpendicular to the boundary containing it.
- The pressure 'inside' a liquid increases in proportion to its depth.

Other important static pressure equations are:

- Absolute pressure = gauge pressure + atmospheric pressure.
- Pressure (p) at depth (h) in a liquid is given by $p = \rho gh$.
- A general equation for a fluid at rest is

$$pdA - \left(p + \frac{dp}{dz}\right)dA - \rho g dA dz = 0$$

This relates to an infinitesimal vertical cylinder of fluid.

5.2 Flow equations

Flow of a fluid may be one dimensional (1D), two dimensional (2D) or three dimensional

The stream tube for conservation of mass

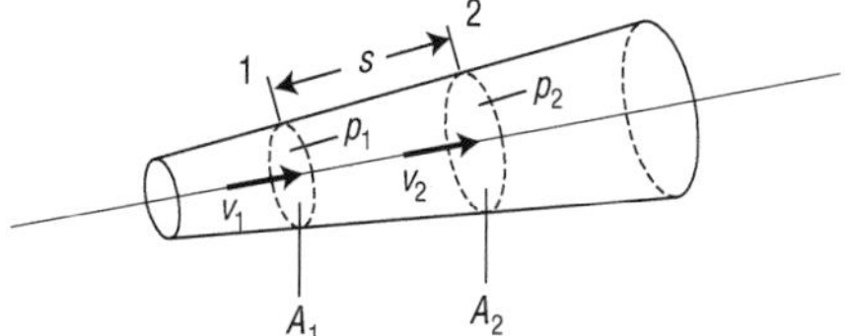

The stream tube and element for the momentum equation

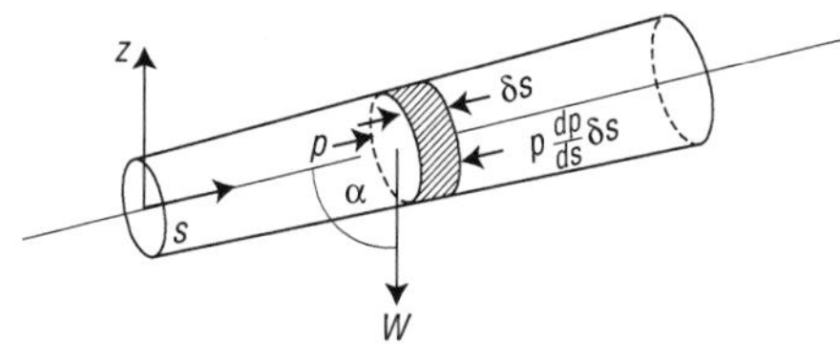

The forces on the element

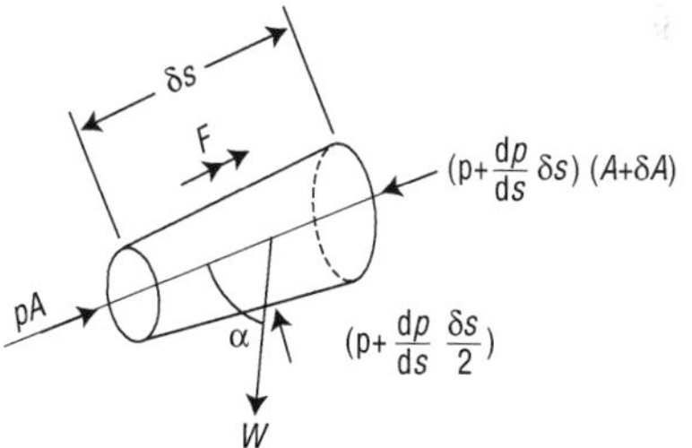

Control volume for the energy equation

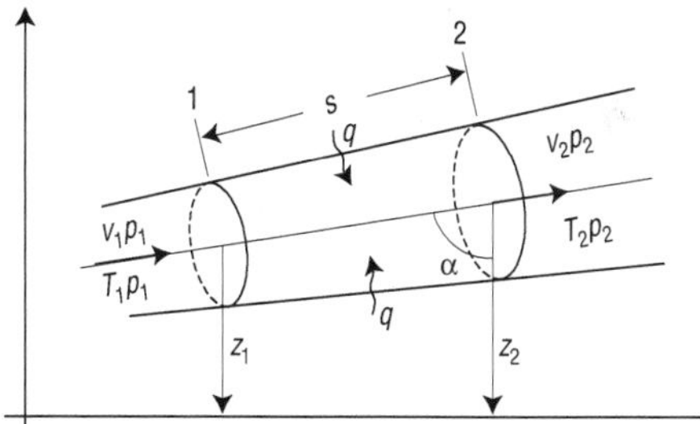

Fig. 5.2 Stream tube/fluid elements: 1-D flow

(3D) depending on the way that the flow is constrained.

5.2.1 1D Flow

1-D flow has a single direction co-ordinate x and a velocity in that direction of u. Flow in a pipe or tube is generally considered one dimensional.

Table 5.2 Fluid principles

Law	*Basis*	*Resulting equations*
Conservation of mass	Matter (in a stream tube or anywhere else) cannot be created or destroyed.	$\rho v A$ = constant
Conservation of momentum	The rate of change of momentum in a given direction = algebraic sum of the forces acting in that direction (Newton's second law of motion).	$\int \sqrt{\frac{dp}{p}} + \frac{1}{2} v^2 + gz$ = constant This is Bernoulli's equation
Conservation of energy	Energy, heat and work are convertible into each other and are in balance in a steadily operating system.	$c_p T + \frac{v^2}{2}$ = constant for an adiabatic (no heat transferred) flow system
Equation of state	Perfect gas state: $p/\rho T = r$ and the first law of thermodynamics	$p = k\rho^{\gamma}$ k = constant γ = ratio of specific heats c_p/c_v

The equations for 1D flow are derived by considering flow along a straight stream tube (see Figure 5.2). Table 5.2 shows the principles, and their resulting equations.

5.2.2 2D Flow

2D flow (as in the space between two parallel flat plates) is that in which all velocities are parallel to a given plane. Either rectangular (x,y) or polar (r, θ) co-ordinates may be used to describe the characteristics of 2D flow. Table 5.3 and Figure 5.3 show the fundamental equations.

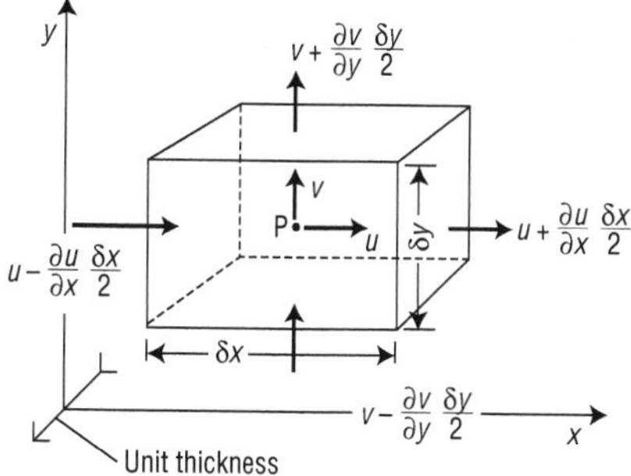

Polar co-ordinates

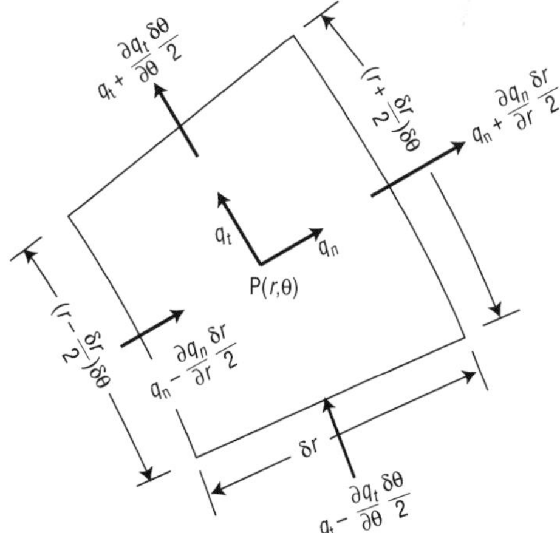

Fig. 5.3 The continuity equation basis in 2-D

Table 5.3 2D flow: fundamental equations

Basis	*The equation*	*Explanation*
Laplace's equation	$\frac{\partial^2 \phi}{\partial x^2} + \frac{\partial^2 \phi}{\partial y^2} = 0 = \frac{\partial^2 \psi}{\partial x^2} + \frac{\partial^2 \psi}{\partial y^2}$ or $\nabla^2 \phi = \nabla^2 \psi = 0$, where $\nabla^2 = \frac{\partial^2}{\partial x^2} + \frac{\partial^2}{\partial y^2}$	A flow described by a unique velocity potential is irrotational.
Equation of motion in 2D	$\frac{\partial u}{\partial t} + u \frac{\partial u}{\partial x} + v \frac{\partial u}{\partial y} = \frac{1}{\rho}\left(X - \frac{\partial p}{\partial x}\right)$ $\frac{\partial v}{\partial t} + u \frac{\partial v}{\partial x} + v \frac{\partial v}{\partial t} = \frac{1}{\rho}\left(Y - \frac{\partial p}{\partial y}\right)$	The principle of force = mass × acceleration (Newton's law of motion) applies to fluids and fluid particles.

Equation of continuity in 2D (incompressible flow)	$\frac{\partial u}{\partial x} + \frac{\partial v}{\partial y} = 0$ or, in polar $\frac{q_n}{r} + \frac{\partial q_n}{\partial r} + \frac{1}{r}\frac{\partial q_t}{\partial \theta} = 0$	If fluid velocity increases in the x direction, it must decrease in the y direction (see Figure 5.3).
Equation of vorticity	$\frac{\partial v}{\partial x} - \frac{\partial u}{\partial y} = \varsigma$ or, in polar: $\varsigma = \frac{q_t}{r} + \frac{\partial q_t}{\partial r} - \frac{1}{r}\frac{\partial q_n}{\partial \theta}$	A rotating or spinning element of fluid can be investigated by assuming it is a solid (see Figure 5.4).
Stream function ψ (incompressible flow)	Velocity at a point is given by: $u = \frac{\partial \psi}{\partial y} \quad v = \frac{\partial \psi}{\partial x}$	ψ is the stream function. Lines of constant ψ give the flow pattern of a fluid stream (see Figure 5.5).
Velocity potential ϕ (irrotational 2D flow)	Velocity at a point is given by: $u = \frac{\partial \phi}{\partial x} \quad v = \frac{\partial \phi}{\partial y}$	ϕ is defined as: $\phi = \int_{op} q \cos \beta \, ds$ (see Figure 5.6).

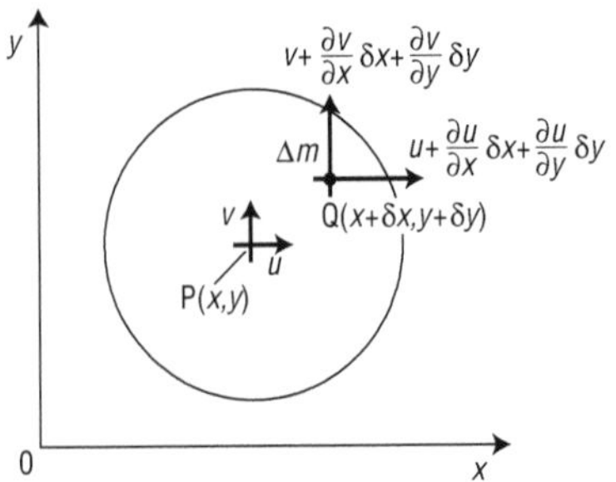

Fig. 5.4 The vorticity equation basis in 2-D

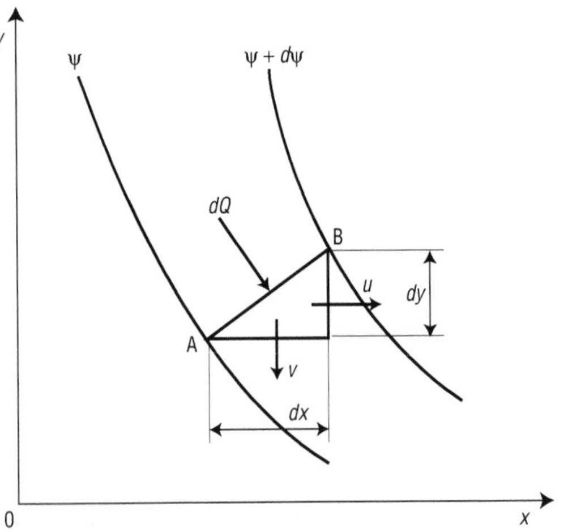

Fig. 5.5 Flow rate (q) and stream function (ψ) relationship

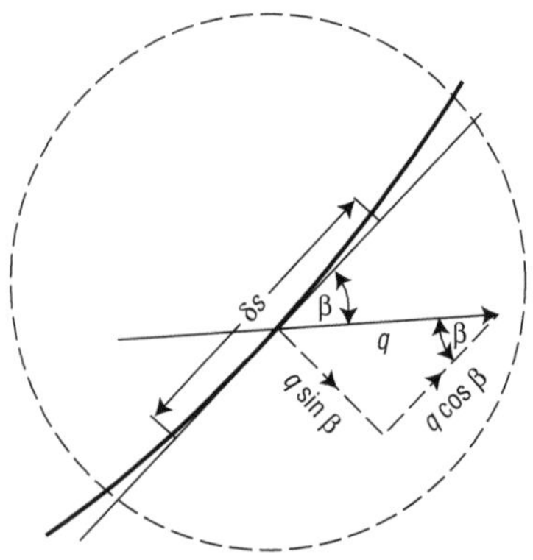

Fig. 5.6 Velocity potential basis

5.2.3 The Navier-Stokes equations

The Navier-Stokes equations are written as:

$$\rho\left(\frac{\partial u}{\partial t}+u\frac{\partial u}{\partial x}+v\frac{\partial u}{\partial y}\right)=\rho X-\frac{\partial p}{\partial x}+\mu\left(\frac{\partial^2 u}{\partial x^2}+\frac{\partial^2 u}{\partial y^2}\right)$$

$$\underbrace{\rho\left(\frac{\partial v}{\partial t}+u\frac{\partial v}{\partial x}+v\frac{\partial v}{\partial y}\right)}_{\text{Inertia term}}=\underbrace{\rho Y}_{\text{Body force term}}-\underbrace{\frac{\partial p}{\partial y}}_{\text{Pressure term}}+\underbrace{\mu\left(\frac{\partial^2 v}{\partial x^2}+\frac{\partial^2 v}{\partial y^2}\right)}_{\text{Viscous term}}$$

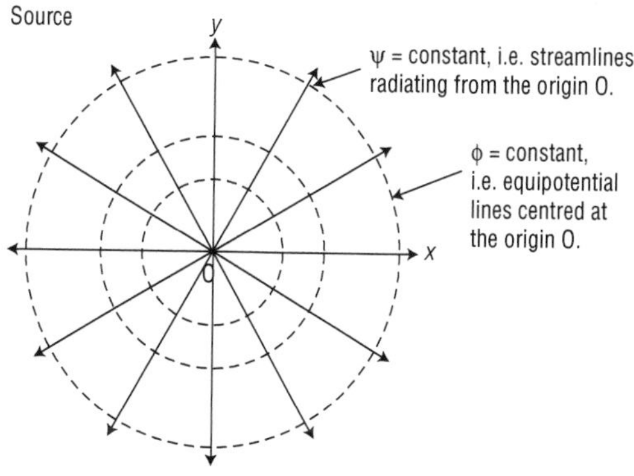

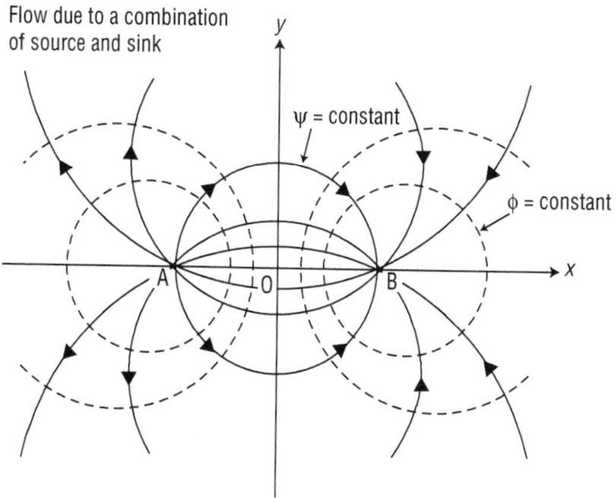

Fig. 5.7 Sources, sinks and combination

5.2.4 Sources and sinks

A *source* is an arrangement where a volume of fluid ($+q$) flows out evenly from an origin toward the periphery of an (imaginary) circle around it. If q is negative, such a point is termed a *sink* (see Figure 5.7). If a source and sink of equal strength have their extremities infinitesimally close to each other, whilst increasing the strength, this is termed a *doublet.*

5.3 Flow regimes

5.3.1 General descriptions

Flow regimes can be generally described as follows (see Figure 5.8):

Steady flow	Flow parameters at any point do not vary with time (even though they may differ between points)
Unsteady flow	Flow parameters at any point vary with time
Laminar flow	Flow which is generally considered smooth, i.e. not broken up by eddies
Turbulent flow	Non-smooth flow in which any small disturbance is magnified, causing eddies and turbulence
Transition flow	The condition lying between laminar and turbulent flow regimes

5.3.2 Reynolds number

Reynolds number is a dimensionless quantity which determines the nature of flow of fluid over a surface.

$$\text{Reynolds number } (Re) = \frac{\text{Inertia forces}}{\text{Viscous forces}}$$

$$= \frac{\rho VD}{\mu} = \frac{VD}{\nu}$$

where ρ = density
μ = dynamic viscosity
ν = kinematic viscosity
V = velocity
D = effective diameter

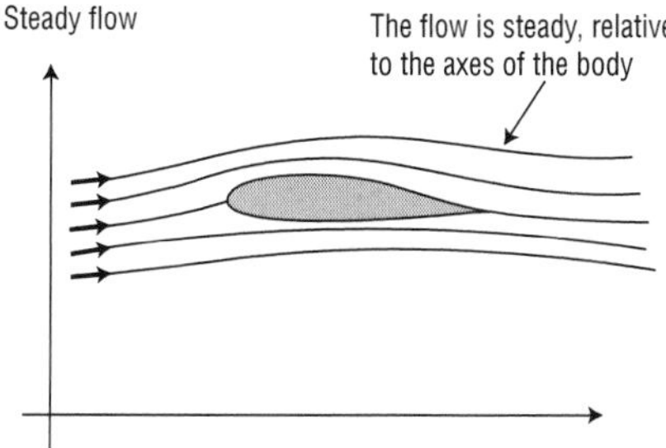

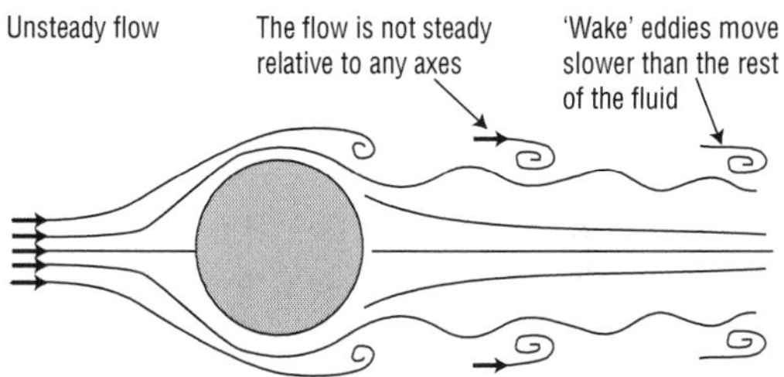

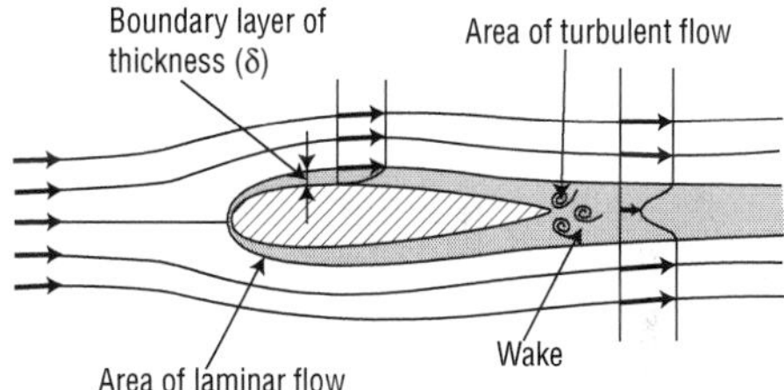

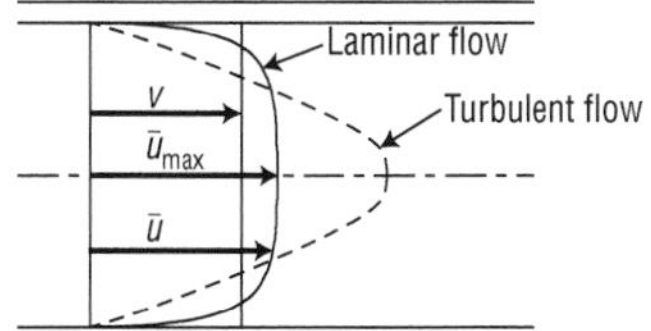

Fig. 5.8 Flow regimes

Low Reynolds numbers (below about 2000) result in laminar flow. High Reynolds numbers (above about 2300) result in turbulent flow.

Values of *Re* for 2000 < *Re* < 2300 are generally considered to result in transition flow. Exact flow regimes are difficult to predict in this region.

5.4 Boundary layers

5.4.1 Definitions

The **boundary layer** is the region near a surface or wall where the movement of the fluid flow is governed by frictional resistance.

The **main flow** is the region outside the boundary layer which is not influenced by frictional resistance and can be assumed to be 'ideal' fluid flow.

Boundary layer thickness: it is convention to assume that the edge of the boundary layer lies at a point in the flow which has a velocity equal to 99% of the local mainstream velocity.

5.4.2 Some boundary layer equations

Figure 5.9 shows boundary layer velocity profiles for dimensional and non-dimensional cases. The non-dimensional case is used to allow comparison between boundary layer profiles of different thickness.

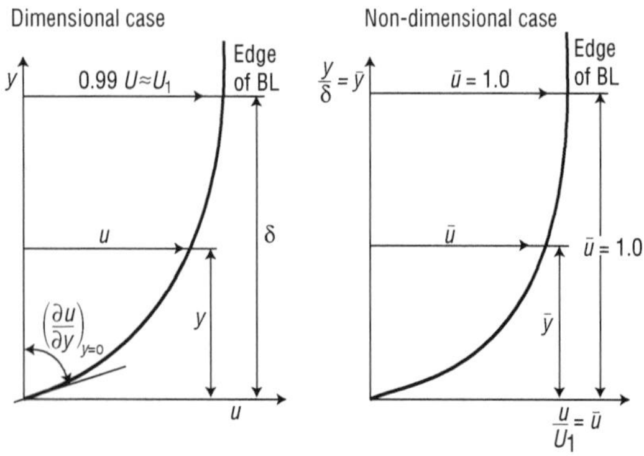

Fig. 5.9 boundary layer velocity profiles

where:

μ = velocity parallel to the surface
y = perpendicular distance from the surface
δ = boundary layer thickness
U_1 = mainstream velocity
$\bar{u}$ = velocity parameters u/U_1 (non-dimensional)
$\bar{y}$ = distance parameter y/δ (non-dimensional)

Boundary layer equations of turbulent flow:

$$\rho\left(\bar{u}\frac{\partial \bar{u}}{\partial x} + \frac{\partial \bar{u}}{\partial y}\right) = -\frac{\partial \bar{p}}{\partial x} + \frac{\partial \tau}{\partial y}$$

$$\tau = \mu\frac{\partial \bar{u}}{\partial y} - \overline{\rho u' v'}$$

$$\frac{\partial \bar{p}}{\partial y} = 0$$

$$\frac{\partial \bar{u}}{\partial x} + \frac{\partial \bar{v}}{\partial y} = 0$$

5.5 Isentropic flow

For flow in a smooth pipe with no abrupt changes of section:

continuity equation $\quad \dfrac{d\rho}{\rho} + \dfrac{du}{u} + \dfrac{dA}{A} = 0$

equation of momentum conservation $\quad -dpA = (A\rho u)du$

isentropic relationship $\quad p = c\rho^k$

sonic velocity $\quad a^2 = \dfrac{dp}{d\rho}$

These lead to an equation being derived on the basis of mass continuity:

i.e. $\quad \dfrac{dp}{\rho} = -M^2\dfrac{du}{u}$

or

$$M^2 = \frac{d\rho}{d\rho}\bigg/\frac{du}{u}$$

Table 5.4 Isentropic flows

Pipe flows	$\dfrac{-dp}{\rho}\Big/\dfrac{du}{u} = M^2$
Convergent nozzle flows	Flow velocity $u = \sqrt{2\left(\dfrac{k}{k-1}\right)\left(\dfrac{p_0}{\rho_0}\right)\left[1 - \dfrac{\rho}{p_0}^{\frac{k-1}{k}}\right]}$ Flow rate $m = \rho u A$
Convergent-divergent nozzle flow	Area ratio $\dfrac{A}{A^*} = \dfrac{\left(\dfrac{2}{k+1}\right)^{\frac{1}{k-1}}\left(\dfrac{p_0}{p}\right)^{1/k}}{\sqrt{\dfrac{k+1}{k-1}\left[1 - \dfrac{p_0}{p}^{\frac{(1-k)}{k}}\right]}}$

Table 5.4 shows equations relating to convergent and convergent-divergent nozzle flow.

5.6 Compressible 1D flow

Basic equations for 1D compressible flow are Euler's equation of motion in the steady state along a streamline:

$$\frac{1}{\rho}\frac{dp}{ds} + \frac{d}{ds}\left(\frac{1}{2}u^2\right) = 0$$

or

$$\int \frac{dp}{\rho} + \frac{1}{2}u^2 = \text{constant}$$

so:

$$\frac{k}{k-1}RT + \frac{1}{2}u^2 = \text{constant}$$

$$\frac{p_0}{p} = \left(\frac{T_0}{T}\right)^{k/(k-1)} = \left(1 + \frac{k-1}{2}M^2\right)^{k/(k-1)}$$

where T_0 = total temperature.

5.7 Normal shock waves

5.7.1 1D flow

A shock wave is a pressure front which travels at speed through a gas. Shock waves cause an increase in pressure, temperature, density and entropy and a decrease in normal velocity.

Equations of state and equations of conservation applied to a unit area of shock wave give (see Figure 5.10):

State $p_1/\rho_1 T_1 = p_2/\rho_2 T_2$

Mass flow $m = \rho_1 u_1 = \rho_2 u_2$

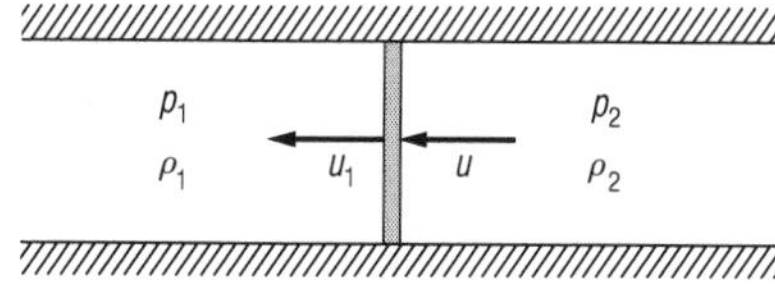

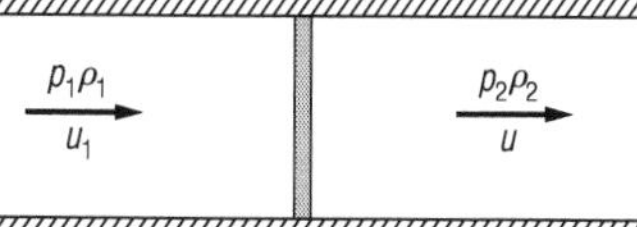

Fig. 5.10(a) 1-D shock waves

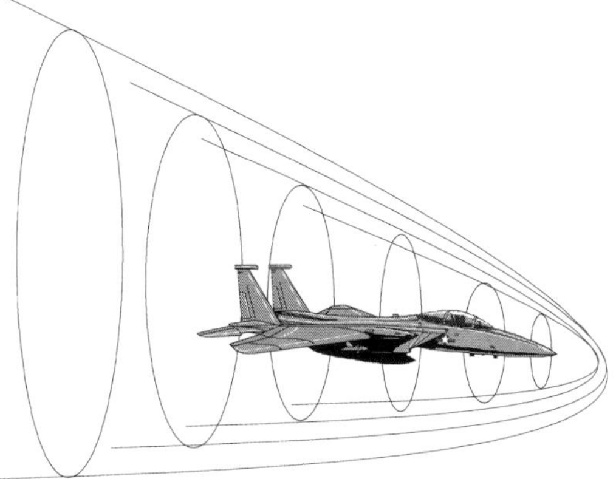

Fig. 5.10(b) Aircraft shock waves

Momentum $p_1 + p_1 u_1^2 = p_2 + \rho_2 u_2^2$

Energy $c_p\, T_1 + \dfrac{u_1^2}{2} = c_p\, T_2 + \dfrac{u_2^2}{2} = c_p\, T_0$

Pressure and density relationships across the shock are given by the Rankine-Hugoniot equations:

$$\frac{p_2}{p_1} = \frac{\dfrac{\gamma+1}{\gamma-1}\dfrac{\rho_2}{\rho_1} - 1}{\dfrac{\gamma+1}{\gamma-1} - \dfrac{\rho_2}{\rho_1}}$$

$$\frac{\rho_2}{\rho_1} = \frac{\dfrac{(\gamma+1)p_2}{(\gamma-1)p_1} + 1}{\dfrac{\gamma+1}{\gamma-1} + \dfrac{p_2}{p_1}}$$

Static pressure ratio across the shock is given by:

$$\frac{p_1}{p_2} = \frac{2\gamma\, M_2^2 - (\gamma-1)}{\gamma+1}$$

Temperature ratio across the shock is given by:

$$\frac{T_2}{T_1} = \frac{p_2}{p_1} \Big/ \frac{\rho_2}{\rho_1}$$

$$\frac{T_2}{T_1} = \left(\frac{2\gamma\, M_1^2 - (\gamma+1)}{\gamma+1}\right)\left(\frac{2 + (\gamma-1)M_1^2}{(\gamma+1)M_1^2}\right)$$

Velocity ratio across the shock is given by:

From continuity: $u_2/u_1 = \rho_1/\rho_2$

so: $$\frac{u_2}{u_1} = \frac{2 + (\gamma-1)M_1^2}{(\gamma+1)M_1^2}$$

In axisymmetric flow the variables are independent of θ so the continuity equation can be expressed as:

$$\frac{1}{R^2}\frac{\partial(R^2\, q_R)}{\partial R} + \frac{1}{R\sin\varphi}\frac{\partial(\sin\varphi q_\varphi)}{\partial\varphi} = 0$$

Similarly in terms of stream function ψ:

$$q_R = \frac{1}{R^2 \sin \varphi} \frac{\partial \psi}{\partial \varphi}$$

$$q_\varphi = \frac{1}{R \sin \psi} \frac{\partial \psi}{\partial R}$$

Additional shock wave data is given in Appendix 5. Figure 5.10(b) shows the practical effect of shock waves as they form around a supersonic aircraft.

5.7.2 The pitot tube equation

An important criterion is the Rayleigh supersonic pitot tube equation (see Figure 5.11).

$$\text{Pressure ratio: } \frac{p_{02}}{p_1} = \left[\frac{\gamma + 1}{2} M_1^2\right]^{\gamma/(\gamma - 1)}$$

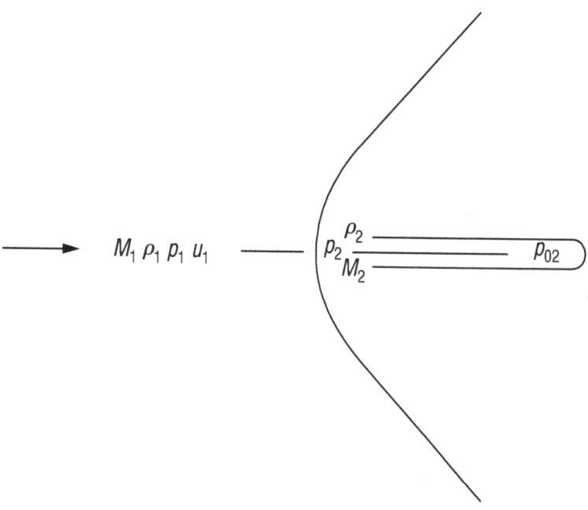

Fig. 5.11 Pitot tube relations

$$\frac{2\gamma M_1^2 - (\gamma - 1)}{\gamma + 1}$$

5.8 Axisymmetric flows

Axisymmetric potential flows occur when bodies such as cones and spheres are aligned

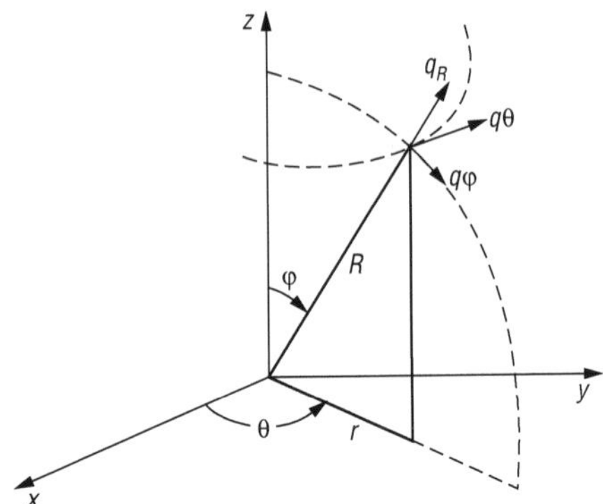

Fig. 5.12 Spherical co-ordinates for axisymmetric flows

into a fluid flow. Figure 5.12 shows the layout of spherical co-ordinates used to analyse these types of flow.

Relationships between the velocity components and potential are given by:

$$q_R = \frac{\partial \phi}{\partial R} \quad q_\theta = \frac{1}{R \sin \varphi} \frac{\partial \phi}{\partial \theta} \quad q_\varphi = \frac{1}{r} \frac{\partial \phi}{\partial \varphi}$$

5.9 Drag coefficients

Figures 5.13(a) and (b) show drag types and 'rule of thumb' coefficient values.

Shape	*Pressure drag* D_P *(%)*	*Friction drag* D_f *(%)*
 U	0	100
 U	≈ 10	≈ 90
 U	≈ 90	≈ 10
 U	100	0

Fig. 5.13(a) Relationship between pressure and fraction drag: 'rule of thumb'

Shape	*Dimensional ratio*	*Datum area, A*	*Approximate drag coefficient, C_D*
Cylinder (flow direction)	$l/d = 1$		0.91
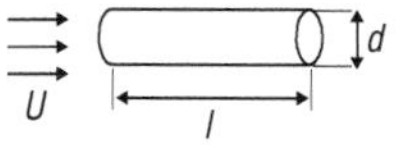	2		0.85
	4	$\frac{\pi}{4} d^2$	0.87
	7		0.99
Cylinder (right angles to flow)	$l/d = 1$		0.63
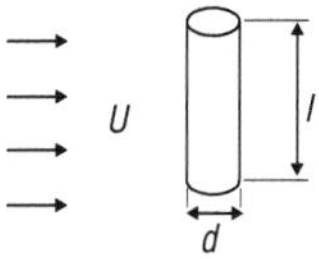	2		0.68
	5		0.74
	10	dl	0.82
	40		0.98
	∞		1.20
Hemisphere (bottomless)	I	$\frac{\pi}{4} d^2$	0.34
	II		1.33
Cone	$a = 60^\circ$	$\frac{\pi}{4} d^2$	0.51
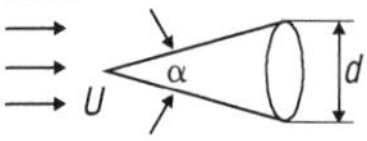	$a = 30^\circ$		0.34
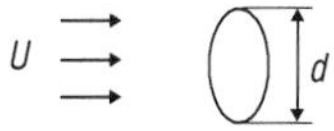		$\frac{\pi}{4} d^2$	1.2
Bluff bodies			
Rough Sphere ($Re = 10^6$)			0.40
Smooth Sphere ($Re = 10^6$)			0.10
Hollow semi-sphere opposite stream			1.42
Hollow semi-sphere facing stream			0.38
Hollow semi-cylinder opposite stream			1.20
Hollow semi-cylinder facing stream			2.30
Squared flat plate at 90°			1.17
Long flat plate at 90°			1.98
Open wheel, rotating, $h/D = 0.28$			0.58
Streamlined bodies			
Laminar flat plate ($Re = 10^6$)			0.001
Turbulent flat plate ($Re = 10^6$)			0.005
Airfoil section, minimum			0.006
Airfoil section, at stall			0.025
2-element airfoil			0.025
4-element airfoil			0.05
Subsonic aircraft wing, minimum			0.05
Subsonic aircraft wing, at stall			0.16
Subsonic aircraft wing, minimum			0.005
Subsonic aircraft wing, at stall			0.09
Aircraft wing (supersonic)			n.a.
Aircraft -general			
Subsonic transport aircraft			0.012
Supersonic fighter, $M = 2.5$			0.016
Airship			0.020–0.025
Helicopter download			0.4–1.2

Fig. 5.13(b) Drag coefficients for standard shapes

Section 6

Basic aerodynamics

6.1 General airfoil theory

When an airfoil is located in an airstream, the flow divides at the leading edge, the stagnation point. The camber of the airfoil section means that the air passing over the top surface has further to travel to reach the trailing edge than that travelling along the lower surface. In accordance with Bernoulli's equation the higher velocity along the upper airfoil surface results in a lower pressure, producing a lift force. The net result of the velocity differences produces an effect equivalent to that of a parallel air stream and a rotational velocity ('vortex') see Figures 6.1 and 6.2.

For the case of a theoretical finite airfoil section, the pressure on the upper and lower surface tries to equalize by flowing round the tips. This rotation persists downstream of the wing resulting in a long U-shaped vortex (see Figure 6.1). The generation of these vortices needs the input of a continuous supply of energy; the net result being to increase the drag of the wing, i.e. by the addition of so-called *induced drag*.

6.2 Airfoil coefficients

Lift, drag and moment (L, D, M) acting on an aircraft wing are expressed by the equations:

$$\text{Lift } (L) \text{ per unit width} = C_L l^2 \frac{\rho U^2}{2}$$

(a)
An effective rotational velocity (vortex) superimposed on the parallel airstream

(b)
Pressures equalize by flows around the tip
Tip
Midspan
Tip

(c)
Finite airfoil
'Horse-shoe' vortex persists downstream
Core of vortex

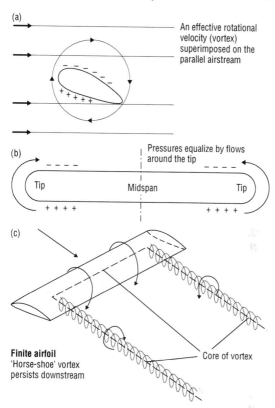

Fig. 6.1 Flows around a finite 3-D airfoil

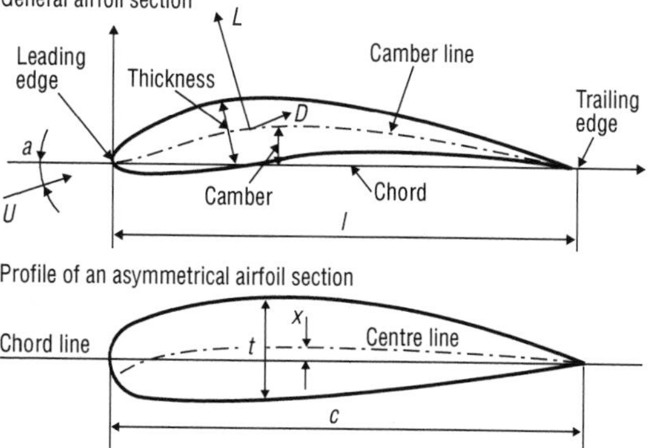

Fig. 6.2 Airfoil sections: general layout

Drag (D) per unit width $= C_D l^2 \dfrac{\rho U^2}{2}$

Moment (M) about LE or

1/4 chord $= C_M l^2 \dfrac{\rho U^2}{2}$

per unit width.

C_L, C_D and C_M are the lift, drag and moment coefficients, respectively. Figure 6.3 shows typical values plotted against the angle of attack, or incidence, (α). The value of C_D is small so a value of 10 C_D is often used for the characteristic curve. C_L rises towards stall point and then falls off dramatically, as the wing enters the stalled condition. C_D rises gradually, increasing dramatically after the stall point. Other general relationships are:

- As a rule of thumb, a Reynolds number of $Re \cong 10^6$ is considered a general flight condition.
- Maximum C_L increases steadily for Reynolds numbers between 10^5 and 10^7.
- C_D decreases rapidly up to Reynolds numbers of about 10^6, beyond which the rate of change reduces.
- Thickness and camber both affect the maximum C_L that can be achieved. As a general rule, C_L increases with thickness and then reduces again as the airfoil becomes even thicker. C_L generally increases as camber increases. The minimum C_D achievable increases fairly steadily with section thickness.

6.3 Pressure distributions

The pressure distribution across an airfoil section varies with the angle of attack (α). Figure 6.4 shows the effect as α increases, and the notation used. The pressure coefficient C_p reduces towards the trailing edge.

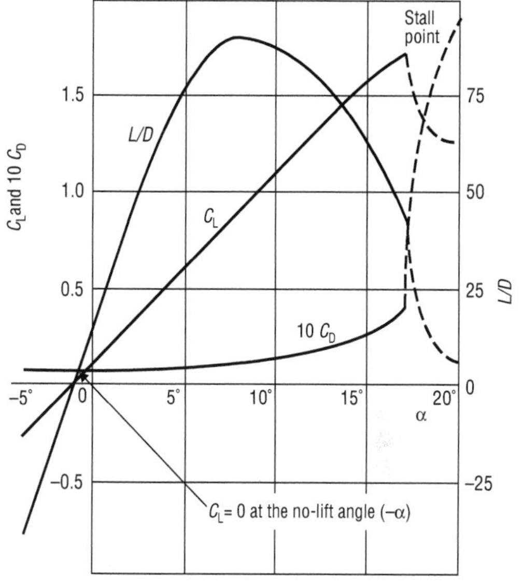

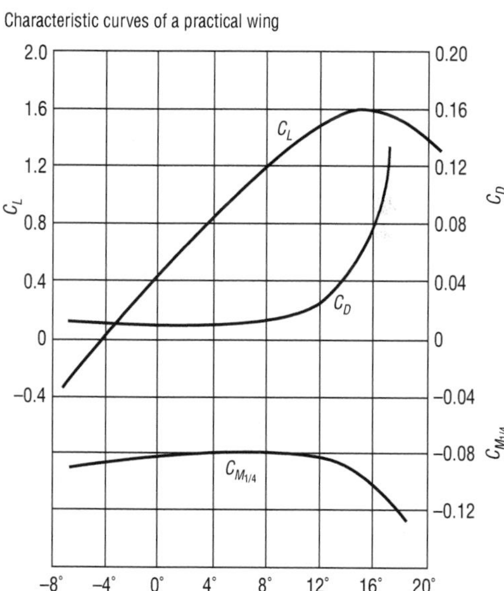

Fig. 6.3 Airfoil coefficients

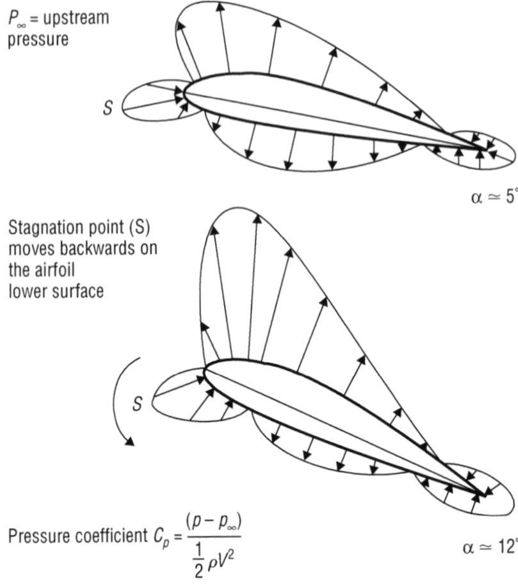

Fig. 6.4 Airfoil pressure coefficient (Cp)

6.4 Aerodynamic centre

The aerodynamic centre (AC) is defined as the point in the section about which the pitching moment coefficient (C_M) is constant, i.e. does not vary with lift coefficient (C_L). Its theoretical positions are indicated in Table 6.1.

Table 6.1 Position of aerodynamic centre

Condition	*Theoretical positon of the AC*
$\alpha < 10°$	At approx. 1/4 chord somewhere near the chord line.
Section with high aspect ratio	At 50% chord.
Flat or curved plate: inviscid, incompressible flow	At approx. 1/4 chord.

Using common approximations, the following equations can be derived:

$$\frac{x_{AC}}{c} = \frac{9}{c} - \frac{d}{dC_L}(C_{Ma})$$

where C_{Ma} = pitching moment coefficient at distance a back from LE

x_{AC} = position of AC back from LE.

c = chord length.

6.5 Centre of pressure

The centre of pressure (CP) is defined as the point in the section about which there is no pitching moment, i.e. the aerodynamic forces on the entire section can be represented by lift and drag forces acting at this point. The CP does not have to lie within the airfoil profile and can change location, depending on the magnitude of the lift coefficient C_L. The CP is conventionally shown at distance k_{CP} back from the section leading edge (see Figure 6.5). Using

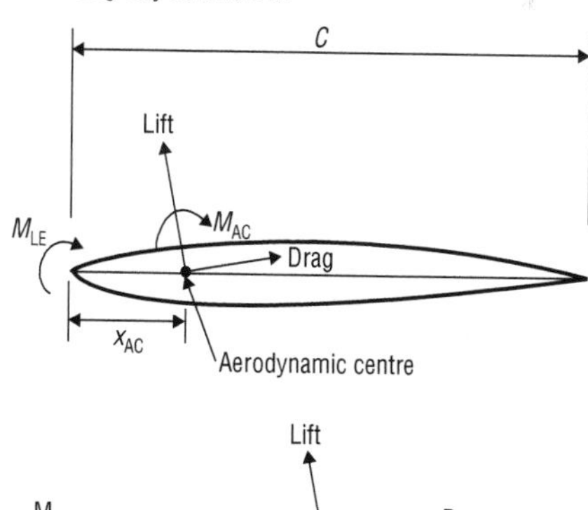

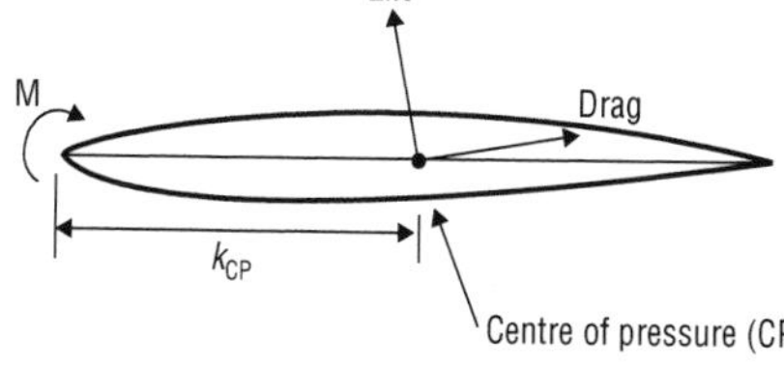

Fig. 6.5 Aerodynamic centre and centre of pressure

the principle of moments the following expression can be derived for k_{CP}:

$$k_{CP} = \frac{x_{AC}}{c} - \frac{C_{M_{AC}}}{C_L \cos \alpha + C_D \sin \alpha}$$

Assuming that $\cos \alpha \cong 1$ and $C_D \sin \alpha \cong 0$ gives:

$$k_{CP} \cong \frac{x_{AC}}{c} - \frac{C_{M_{AC}}}{C_L}$$

6.6 Supersonic conditions

As an aircraft is accelerated to approach supersonic speed the equations of motion which describe the flow change in character. In order to predict the behaviour of airfoil sections in upper subsonic and supersonic regions, compressible flow equations are required.

6.6.1 Basic definitions

- M Mach number
- M_∞ Free stream Mach number
- M_c Critical Mach number, i.e. the value of which results in flow of $M_\infty = 1$ at some location on the airfoil surface.

Figure 6.6 shows approximate forms of the pressure distribution on a two-dimensional airfoil around the critical region. Owing to the complex non-linear form of the equations of motion which describe high speed flow, two popular simplifications are used: the *small perturbation* approximation and the so-called *exact* approximation.

6.6.2 Supersonic effects on drag

In the supersonic region, induced drag (due to lift) increases in relation to the parameter $\sqrt{M^2 - 1}$ function of the plan form geometry of the wing.

6.6.3 Supersonic effects on aerodynamic centre

Figure 6.7 shows the location of wing aerodynamic centre for several values of tip chord/root chord ratio (γ). These are empirically based results which can be used as a 'rule of thumb'.

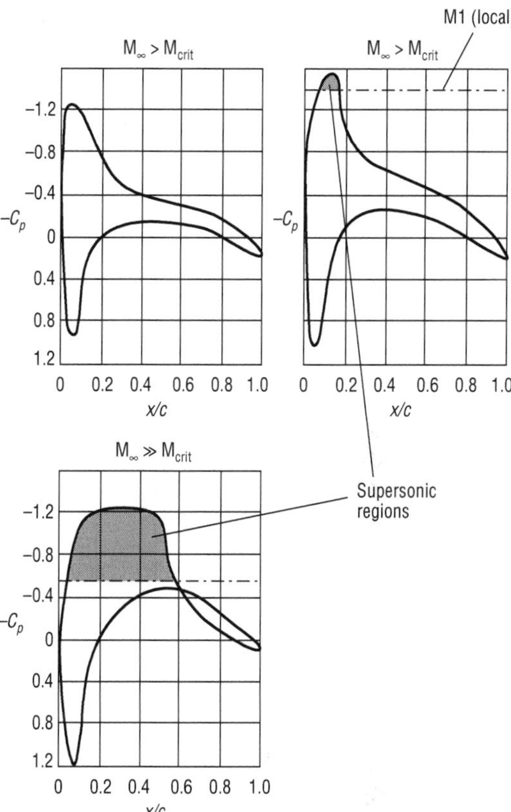

Fig. 6.6 Variation of pressure deterioration (2-D airfoil)

6.7 Wing loading: semi-ellipse assumption

The simplest general loading condition assumption for symmetric flight is that of the semi-ellipse. The equivalent equations for lift, downwash and induced drag become:

For lift:

$$L = \rho \frac{VK_0 \pi s}{2}$$

replacing L by $C_L \, {}^1\!/_2 \rho V^2 S$ gives:

$$K_0 = \frac{C_L VS}{\pi s}$$

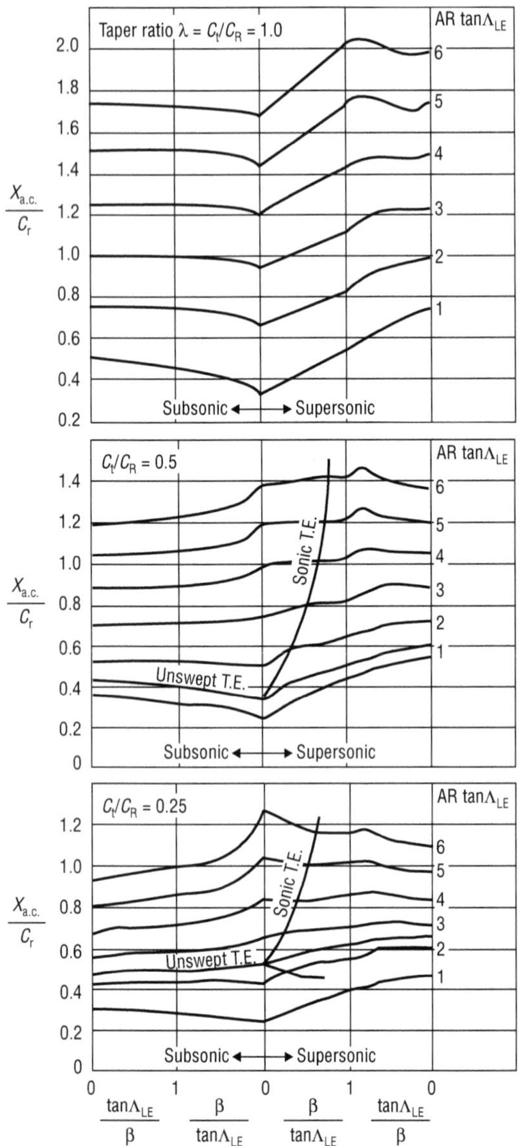

Fig. 6.7 Wing aerodynamic centre location: subsonic/supersonic flight. Originally published in *The AIAA Aerospace Engineers Design Guide*, 4th Edition. Copyright © 1998 by The American Institute of Aeronautics and Astronautics Inc. Reprinted with permission.

For downwash velocity (w):

$$w = \frac{K_0}{4S}, \text{ i.e. it is constant along the span.}$$

For induced drag (vortex):

$$D_{D_V} = \frac{C_L{}^2}{\pi \mathrm{AR}}$$

$$\text{where aspect ratio (AR)} = \frac{\text{span}^2}{\text{area}} = \frac{4s^2}{S}$$

Hence, C_{D_V} falls (theoretically) to zero as aspect ratio increases. At zero lift in symmetric flight, $C_{D_V} = 0$.

Section 7

Principles of flight dynamics

7.1 Flight dynamics – conceptual breakdown

Flight dynamics is a multi-disciplinary subject consisting of a framework of fundamental mathematical and physical relationships. Figure 7.1 shows a conceptual breakdown of the subject relationships. A central tenet of the framework are the equations of motion, which provide a mathematical description of the physical response of an aircraft to its controls.

7.2 Axes notation

Motions can only be properly described in relation to a chosen system of axes. Two of the most common systems are *earth axes* and *aircraft body axes.*

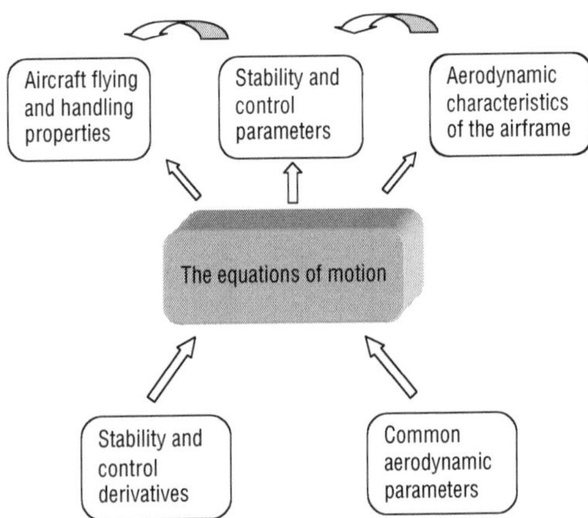

Fig. 7.1 Flight dynamics – the conceptual breakdown

Conventional earth axes are used as a reference frame for 'short-term' aircraft motion.

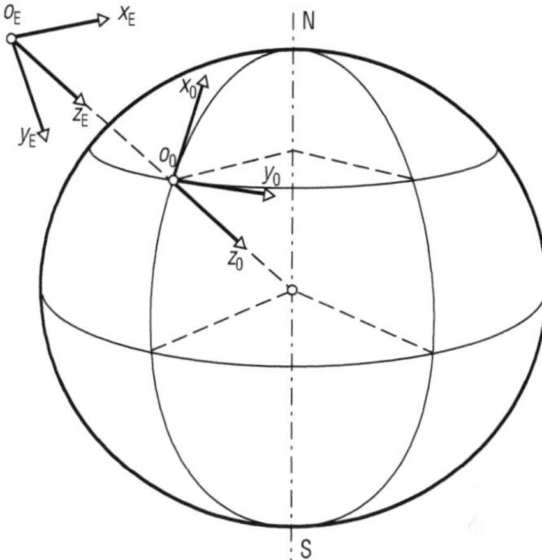

- The horizontal plane o_E, x_E, y_E, lies parallel to the plane o_0, x_0, y_0, on the earth's surface.
- The axis o_E, z_E, points vertically downwards.

Fig. 7.2 Conventional earth axes

7.2.1 Earth axes

Aircraft motion is measured with reference to a fixed earth framework (see Figure 7.2). The system assumes that the earth is flat, an assumption which is adequate for short distance flights.

7.2.2 Aircraft body axes

Aircraft motion is measured with reference to an orthogonal axes system (Ox_b, y_b, z_b) fixed on the aircraft, i.e. the axes move as the aircraft moves (see Figure 7.3).

7.2.3 Wind or 'stability' axes

This is similar to section 7.2.2 in that the axes system is fixed in the aircraft, but with the Ox-axis orientated parallel to the velocity vector V_0 (see Figure 7.3).

7.2.4 Motion variables

The important motion and 'perturbation' variables are force, moment, linear velocity,

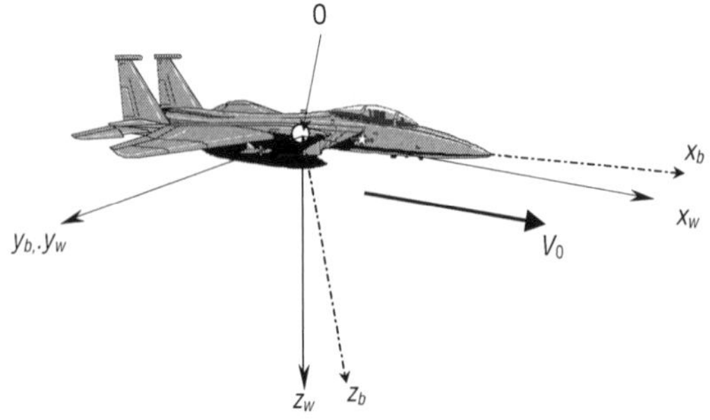

Fig. 7.3 Aircraft body axes

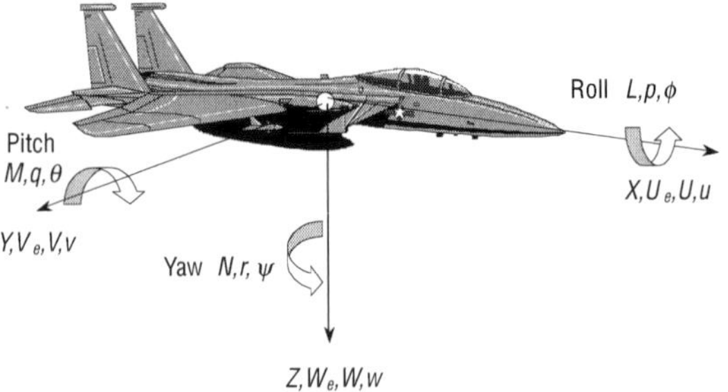

Fig. 7.4 Motion variables: common notation

angular velocity and attitude. Figure 7.4 and Table 7.1 show the common notation used.

7.2.5 Axes transformation

It is possible to connect between axes references: e.g. if Ox_0, y_0, z_0 are wind axes and components in body axes and ϕ, θ, ψ are the angles with respect to each other in roll, pitch and yaw, it can be shown that for linear quantities in matrix format:

$$\begin{bmatrix} Ox_3 \\ Oy_3 \\ Oz_3 \end{bmatrix} = \mathbf{D} \begin{bmatrix} Ox_0 \\ Oy_0 \\ Oz_0 \end{bmatrix}$$

Table 7.1 Motion and perturbation notation

Perturbations			
Aircraft axis	Ox	Oy	Oz
Force	X	Y	Z
Moment	L	M	N
Linear velocity	U	V	W
Angular velocity	p	q	r
Attitude	ϕ	θ	ψ

Motions

X	Axial 'drag' force
Y	Side force
Z	Normal 'lift' force
L	Rolling moment
M	Pitching moment
N	Yawing moment
p	Roll rate
q	Pitch rate
r	Yaw rate
U	Axial velocity
V	Lateral velocity
W	Normal velocity

Where the direction cosine matrix **D** is given by:

$$D = \begin{bmatrix} \cos\theta\cos\psi & \cos\theta\cos\psi & -\sin\theta \\ \sin\phi\sin\theta\cos\psi - \cos\phi\sin\psi & \sin\phi\sin\theta\sin\psi + \cos\phi\sin\psi & \sin\phi\cos\theta \\ \cos\phi\sin\theta\cos\psi + \sin\phi\sin\psi & \cos\phi\sin\theta\cos\psi - \sin\phi\cos\psi & \cos\phi\cos\theta \end{bmatrix}$$

Angular velocity transformations can be expressed as:

$$\begin{bmatrix} p \\ q \\ r \end{bmatrix} = \begin{bmatrix} 1 & 0 & -\sin\theta \\ 0 & \cos\phi & \sin\phi\cos\theta \\ 0 & -\sin\phi & \cos\phi\cos\theta \end{bmatrix} \begin{bmatrix} \dot{\phi} \\ \dot{\theta} \\ \dot{\psi} \end{bmatrix}$$

where p, q, r are angular body rates:

Roll rate $p = \dot{\phi} - \dot{\psi}\sin\theta$

Pitch rate $q = \dot{\theta}\cos\phi + \dot{\psi}\sin\phi\cos\theta$

Yaw rate $r = \dot{\psi}\cos\phi\cos\theta - \dot{\theta}\sin\phi$

where $\dot{\phi}$, $\dot{\theta}$, $\dot{\psi}$ are attitude rates with respect to datum axes

Inverting gives:

$$\begin{bmatrix} \dot{\phi} \\ \dot{\theta} \\ \dot{\psi} \end{bmatrix} = \begin{bmatrix} 1 & \sin\phi \tan\theta & \cos\phi \tan\theta \\ 0 & \cos\phi & -\sin\phi \\ 0 & \sin\phi \sec\theta & \cos\phi \sec\theta \end{bmatrix} \begin{bmatrix} p \\ q \\ r \end{bmatrix}$$

7.3 The generalized force equations

The equations of motions for a rigid aircraft are derived from Newton's second law ($F = ma$) expressed for six degrees of freedom.

7.3.1 Inertial acceleration components

To apply $F = ma$, it is first necessary to define acceleration components with respect to earth ('inertial') axes. The equations are:

$$\left.\begin{aligned} a^{l}{}_{x} &= U - rV + qW - x(q^2 + r^2) + y(pq - r) \\ &\quad + z(pr + q) \\ a^{l}{}_{y} &= V - pW + rU + x(pq + r) - y(p^2 + r^2) \\ &\quad + x(qr - p) \\ a^{l}{}_{z} &= W - qU + pV = x(pr - q) + y(qr + p) \\ &\quad - z(p^2 + q^2) \end{aligned}\right\}$$

where: $a^{l}{}_{x}$, $a^{l}{}_{y}$, $a^{l}{}_{z}$ are vertical acceleration components of a point p(x, y, z) in the rigid aircraft.

U, V, W are components of velocity along the axes Ox, Oy, Oz.

p, q, r are components of angular velocity.

7.3.2 Generalized force equations

The generalized force equations of a rigid body (describing the motion of its centre of gravity) are:

$$\left.\begin{aligned} m(U - rV + qW) &= X \\ m(V - pW + rU) &= Y \\ m(W - qU + pV) &= Z \end{aligned}\right\} \text{ where } m \text{ is the total mass of the body}$$

7.4 The generalized moment equations

A consideration of moments of forces acting at a point p(x, y, z) in a rigid body can be expressed as follows:

Moments of inertia

$I_x = \Sigma\delta m(y^2 + z^2)$	Moment of inertia about Ox axis
$I_y = \Sigma\delta m(x^2 + z^2)$	Moment of inertia about Oy axis
$I_z = \Sigma\delta m(x^2 + y^2)$	Moment of inertia about Oz axis
$I_{xy} = \Sigma\delta m\ xy$	Product of inertia about Ox and Oy axes
$I_{xz} = \Sigma\delta m\ xz$	Product of inertia about Ox and Oz axes
$I_{yz} = \Sigma\delta m\ yz$	Product of inertia about Oy and Oz axes

The simplified moment equations become

$$\left.\begin{array}{l} I_x\dot{p} - (I_y - I_z)\,qr - I_{xz}\,(pq + \dot{r}) = L \\ I_y\dot{q} - (I_x - I_z)\,pr - I_{xz}\,(p^2 - r^2) = M \\ I_z\dot{r} - (I_x - I_y)\,pq - I_{xz}\,(qr + \dot{p}) = N \end{array}\right\}$$

7.5 Non-linear equations of motion

The generalized motion of an aircraft can be expressed by the following set of *non-linear equations of motion*:

$$\left.\begin{array}{l} m(\dot{U} - rV + qW) = X_a + X_g + X_c + X_p + X_d \\ m(\dot{V} - pW + rU) = Y_a + Y_g + Y_c + Y_p + Y_d \\ m(\dot{W} - qU + pV) = Z_a + Z_g + Z_c + Z_p + Z_d \\ I_x\,\dot{p} - (I_y - I_x)\,qr - I_{xz}\,(pq + \dot{r}) = L_a + L_g + \\ L_c + L_p + L_d \\ I_y\,\dot{q} + (I_x - I_z)\,pr + I_{xz}\,(p^2 - r^2) = M_a + M_g + \\ M_c + M_p + M_d \\ I_z\,\dot{r} - (I_x - I_y)\,pq + I_{xz}\,(qr - \dot{p}) = N_a + N_g + \\ N_c + N_p + N_d \end{array}\right\}$$

7.6 The linearized equations of motion

In order to use them for practical analysis, the equations of motions are expressed in their linearized form by using the assumption that all perturbations of an aircraft are small, and about the 'steady trim' condition. Hence the equations become:

$$\left.\begin{array}{l} m(u + qW_e) = X_a + X_g + X_c + X_p \\ m(v + pW_e + rU_e) = Y_a + Y_g + Y_c + Y_p \\ m(w + qU_e) = Z_a + Z_g + Z_c + Z_p \\ I_x\,p - I_{xz}\,r = L_a + L_g + L_c + L_p \\ I_y\,q = M_a + M_g + M_c + M_p \\ I_z\,r - I_{xz}\,p = N_a + N_g + N_c + N_p \end{array}\right\}$$

A better analysis is obtained by substituting appropriate expressions for aerodynamic, gravitational, control and thrust terms. This gives a set of six simultaneous linear differential equations which describe the transient response of an aircraft to small disturbances about its trim condition, i.e.:

$$mu - \mathring{X}_u\,u - \mathring{X}_v\,v - \mathring{X}_w\,w - \mathring{X}_w\,w$$

$$-\mathring{X}_p\,p - (\mathring{X}_q - mW_e)_q - \mathring{X}_r\,r + mg\theta\cos\theta_e = \mathring{X}_\xi\,\xi + \mathring{X}_\eta\,\eta + \mathring{X}_\zeta\,\zeta + \mathring{X}_\tau\,\tau$$

$$-\mathring{Y}_u\,u + mv - \mathring{Y}_v\,v - \mathring{Y}_w\,w - \mathring{Y}_w\,w - (\mathring{Y}_p + mW_e)p$$

$$-\mathring{Y}_q\,q - (\mathring{Y}_r - mU_e)r - mg\phi\cos\theta_e - mg\psi\sin\theta_e = \mathring{Y}_\xi\,\xi + \mathring{Y}_\eta\,\eta = \mathring{Y}_\zeta\,\zeta + \mathring{Y}_\tau\,\tau$$

$$-\mathring{Z}_u\,u - \mathring{Z}_v\,v + (m - \mathring{Z}_w\,w)\,w - \mathring{Z}_w\,w$$

$$-\mathring{Z}_p\,p - (\mathring{Z}_q - mU_e)_q - \mathring{Z}_r\,r + mg\theta\sin\theta_e = \mathring{Z}_\xi\,\xi + \mathring{Z}_\eta\,\eta = \mathring{Z}_\zeta\,\zeta + \mathring{Z}_\tau\,\tau$$

$$-\mathring{L}_u\,u - \mathring{L}_v\,v - \mathring{L}_w\,w - \mathring{L}_w\,w$$

$$+I_x\,p - \mathring{L}_p\,p - \mathring{L}_q\,q - I_{xz}\,r - \mathring{L}_r\,r = \mathring{L}_\xi\,\xi + \mathring{L}_\eta\,\eta = \mathring{L}_\zeta\,\zeta + \mathring{L}_\tau\,\tau$$

$$-\mathring{M}_u\,u - \mathring{M}_v\,v - \mathring{M}_w\,w$$

$$-\mathring{M}_w\,w - \mathring{M}_p\,p - + I_y\,q - \mathring{M}_q\,q - \mathring{M}_r\,r = \mathring{M}_\xi\,\xi + \mathring{M}_\eta\,\eta = \mathring{M}_\zeta\,\zeta + \mathring{M}_\tau\,\tau$$

$$-\mathring{N}_u\,u - \mathring{N}_v\,v - \mathring{N}_w\,w - \mathring{N}_w\,w$$

$$I_{xz}\,p - \mathring{N}_p\,p - \mathring{N}_q\,q + I_z\,r - \mathring{N}_r\,r = \mathring{N}_\xi\,\xi + \mathring{N}_\eta\,\eta = \mathring{N}_\zeta\,\zeta + \mathring{N}_\tau\,\tau$$

Table 7.2 Stability terms

Term	*Meaning*
Static stability	The tendency of an aircraft to converge back to its equilibrium condition after a small disturbance from trim.
Lateral static stability	The tendency of an aircraft to maintain its wings level in the roll direction.
Directional static stability	The tendency of an aircraft to 'weathercock' into the wind to maintain directional equilibrium.
Dynamic stability	The transient motion involved in recovering equilibrium after a small disturbance from trim.
Degree of stability	A parameter expressed by reference to the magnitude of the slope of the $C_m - \alpha$, $C_1 - \phi$ and $C_n - \beta$ characteristics.
Stability margin	The amount of stability in excess of zero or neutral stability.
Stability reversal	Change in sign of pitching moment coefficient (C_m) at high values of lift coefficient (C_L). The result is an unstable pitch-up characteristic (see Figures 7.6 and 7.7).
'Controls fixed' stability	Stability of an aircraft in the condition with its flying control surfaces held at a constant setting for the prevailing trim condition.
'Controls free' stability	Stability of an aircraft in the condition with its flying control surfaces (elevator) free to float at an angle corresponding to the prevailing trim condition.

7.7 Stability

Stability is about the nature of motion of an aircraft after a disturbance. When limited by the assumptions of the linearized equations of motion it is restricted to the study of the motion after a small disturbance about the trim condition. Under linear system assumptions, stability is independent of the character of the disturbing force. In practice, many aircraft display distinctly non-linear characteristics. Some useful definitions are given in Table 7.2, see also Figures 7.5 and 7.6

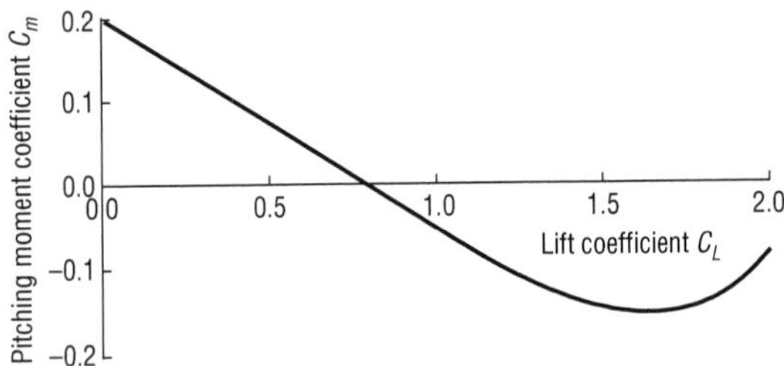

Fig. 7.5 Stability reversal at high lift coefficient

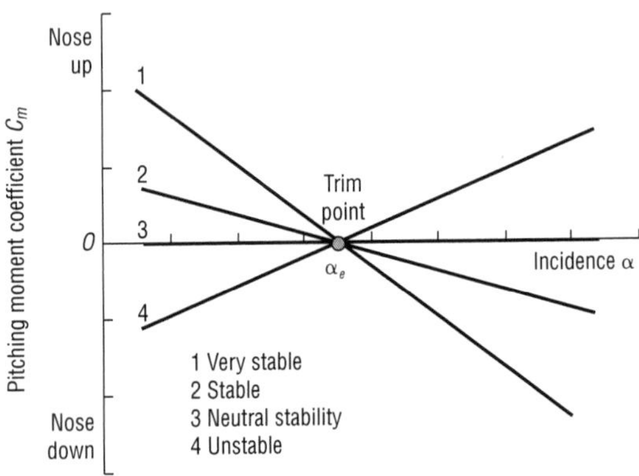

Fig. 7.6 Degree of stability (static, longitudinal)

Section 8

Principles of propulsion

8.1 Propellers

A propeller or airscrew converts the torque of an engine (piston engine or turboprop) into thrust. Propeller blades have an airfoil section which becomes more 'circular' towards the hub. The torque of a rotating propeller imparts a rotational motion to the air flowing through it. Pressure is reduced in front of the blades and increased behind them, creating a rotating slipstream. Large masses of air pass through the propeller, but the velocity rise is small compared to that in turbojet and turbofan engines.

8.1.1 Blade element design theory

Basic design theory considers each section of the propeller as a rotating airfoil. The flow over the blade is assumed to be two dimensional (i.e. no radial component). From Figure 8.1 the following equations can be expressed:

Pitch angle $\phi = \tan^{-1}(V_0/\pi nd)$

The propulsion efficiency of the blade element, i.e. the *blading efficiency*, is defined by:

$$\eta_b = \frac{V_0 dF}{u dQ} = \frac{\tan\phi}{\tan(\phi + \gamma)} = \frac{L/D - \tan\phi}{L/D + \cot\phi}$$

where
u = velocity of blade element = $2\pi nr$
D = drag
L = lift
dF = thrust force acting on blade element
dQ = corresponding torque force
r = radius

Vector diagram for a blade element of a propeller

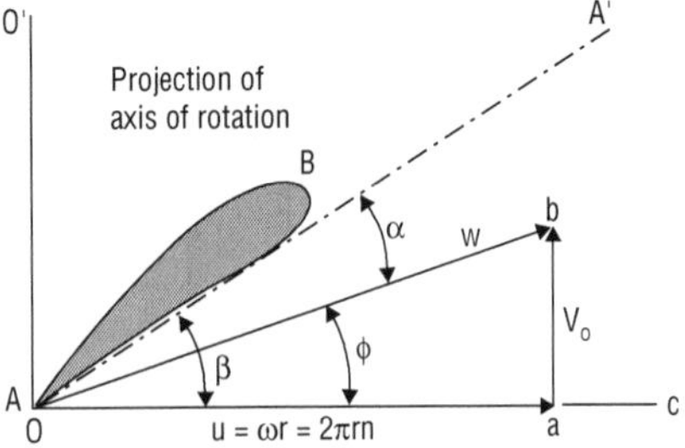

Aerodynamic forces acting on a blade element

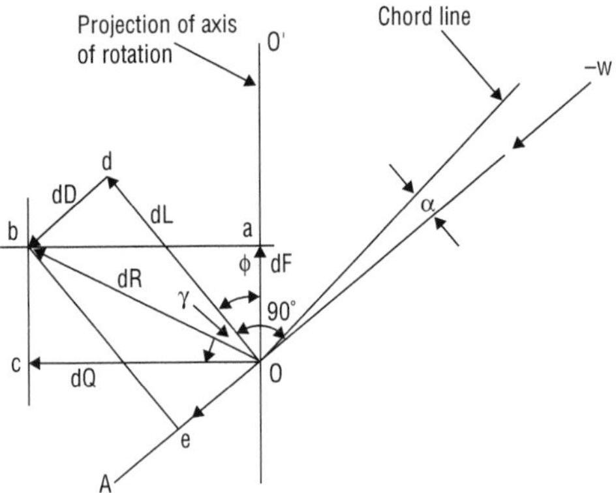

Fig. 8.1 Propeller blade elements

The value of ϕ which makes η_b a maximum is termed the optimum advance angle ϕ_{opt}.

Maximum blade efficiency is given by:

$$(\eta_b)_{\max} = \frac{2\gamma - 1}{2\gamma + 1} = \frac{2(L/D) - 1}{2(L/D) + 1}$$

8.1.2 Performance characteristics

The pitch and angle ϕ have different values at different radii along a propeller blade. It is common to refer to all parameters determining the overall characteristics of a propeller to their values at either $0.7r$ or $0.75r$.

Lift coefficient C_L is a linear function of the angle of attack (α) up to the point where the

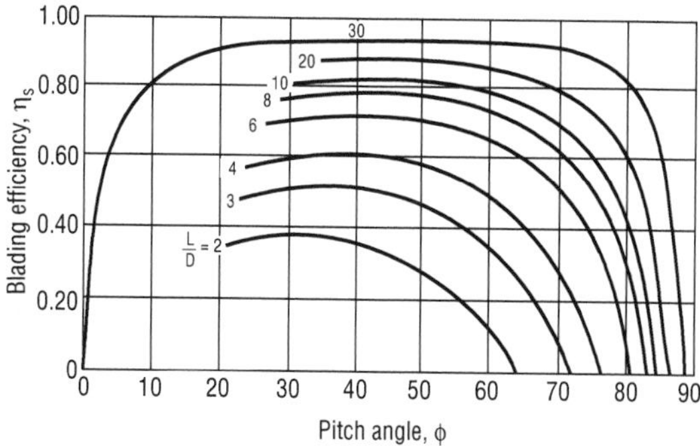

Fig. 8.2 Propeller parameter relationship

blade stalls whilst drag coefficient C_D is quadratic function of α. Figure 8.2 shows broad relationships between blading efficiency, pitch angle and L/D ratio.

8.1.3 Propeller coefficients

It can be shown, neglecting the compressibility of the air, that:

$$f(V_0, n, d_p, \rho, F) = 0$$

Using dimensional analysis, the following coefficients are obtained for expressing the performances of propellers having the same geometry:

$$F = \rho n^2 d^4{}_p C_F \quad Q = \rho n^2 d^5{}_p C_Q \quad P = \rho n^3 d^5{}_p C_p$$

C_F, C_Q and C_P are termed the thrust, torque, and power coefficients. These are normally expressed in USCS units, i.e.:

$$\text{Thrust coefficient } C_F = \frac{F}{\rho n^2 d^4}$$

$$\text{Torque coefficient } C_Q = \frac{Q}{\rho n^2 d^5}$$

$$\text{Power coefficient } C_P = \frac{P}{\rho n^3 d^4}$$

where d = propeller diameter (ft)
n = speed in revs per second
Q = torque (ft lb)
F = thrust (lbf)
P = power (ft lb/s)
ρ = air density (lb s²/ft⁴)

8.1.4 Activity factor

Activity factor (AF) is a measure of the power-absorbing capabilities of a propeller, and hence a measure of its 'solidity'. It is defined as:

$$\mathrm{AF} = \frac{100\,000}{16} \int_{r_h/R}^{r/R=1} \frac{c}{d_P} \left(\frac{r}{R}\right)^3 d\left(\frac{r}{R}\right)$$

8.1.5 Propeller mechanical design

Propeller blades are subjected to:

- Tensile stress due to centrifugal forces.
- Steady bending stress due to thrust and torque forces.
- Bending stress caused by vibration.

Vibration-induced stresses are the most serious hence propellers are designed so that their first order natural reasonant frequency lies above expected operating speeds. To minimize the chance of failures, blades are designed using fatigue strength criteria. Steel blades are often hollow whereas aluminium alloy ones are normally solid.

8.2 The gas turbine engine: general principles

Although there are many variants of gas turbine-based aero engines, they operate using similar principles. Air is compressed by an axial flow or centrifugal compressor. The highly compressed air then passes to a combustion chamber where it is mixed with fuel and ignited. The mixture of air and combustion products expands into the turbine stage which in turn provides the power through a coupling shaft to drive the compressor. The expanding gases then pass out through the engine tailpipe, providing thrust, or can be passed through a further turbine stage to drive a propeller or helicopter rotor. For aeronautical applications the two most important criteria in engine choice are thrust (or power) and specific fuel consumption. Figure 8.3 shows an outline of

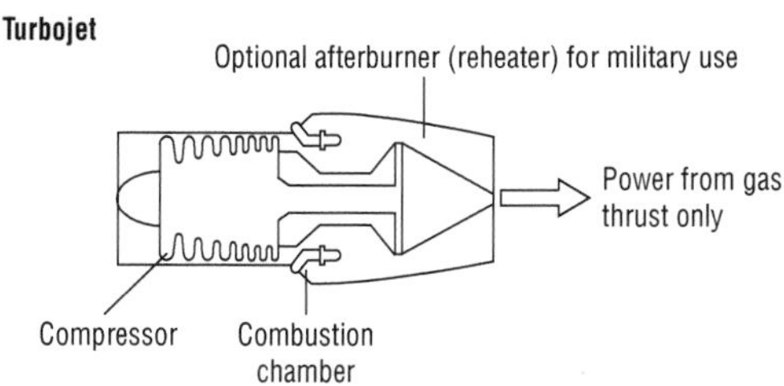

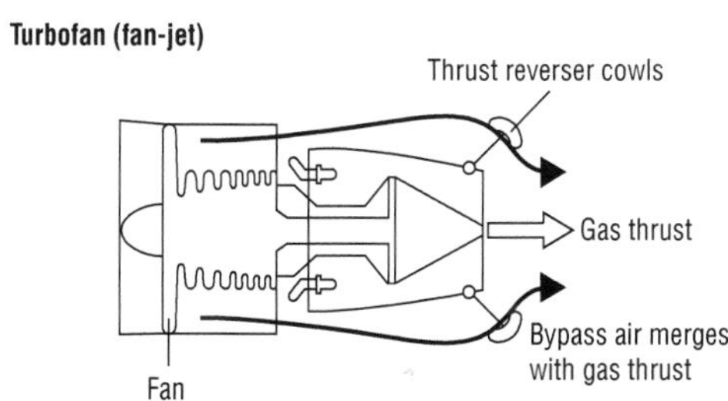

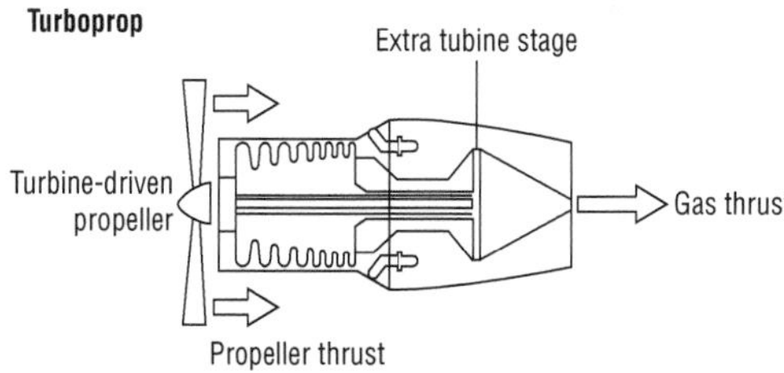

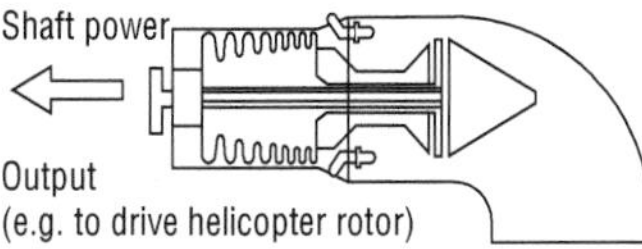

Fig. 8.3 Gas turbine engine types

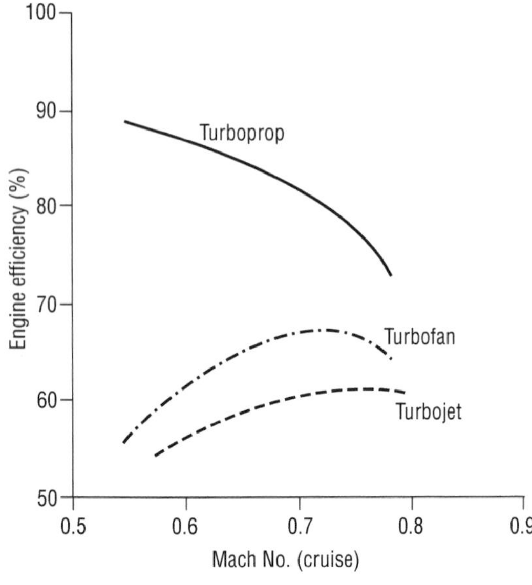

Fig. 8.4 'Order of magnitude' engine efficiencies

the main types and Figure 8.4 an indication of engine efficiency at various flight speeds.

8.2.1 The simple turbojet

The simple turbojet derives all its thrust from the exit velocity of the exhaust gas. It has no separate propeller or 'power' turbine stage. Performance parameters are outlined in Figure 8.5. Turbojets have poor fuel economy and high exhaust noise. The fact that all the air passes through the engine core (i.e. there is no bypass) is responsible for the low propulsive efficiency, except at very high aircraft speed. The Concorde supersonic transport (SST) aircraft is virtually the only commercial airliner that still uses the turbojet. By making the convenient assumption of neglecting Reynolds number, the variables governing the performance of a simple turbojet can be grouped as shown in Table 8.1.

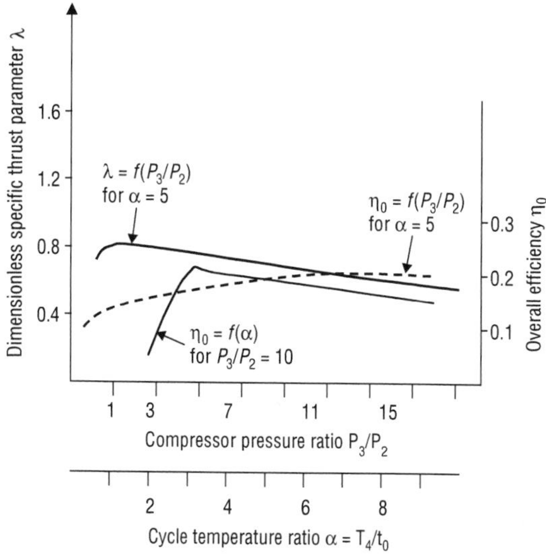

Fig. 8.5 Turbojet performance indicative design points

Table 8.1 Turbojet performance parameter groupings

Non-dimensional group	*Uncorrected*	*Corrected*
Flight speed	$V_0/\sqrt{t_0}$	$V_0\sqrt{\theta}$
Rpm	$N/\sqrt{T}$	$N/\sqrt{\theta}$
Air flow rate	$\dot{W}_a/\sqrt{T/D^2P}$	$\dot{W}_a/\sqrt{\theta/\delta}$
Thrust	F/D^2P	F/δ
Fuel flow rate	$\dot{W}_f J\Delta H_c/D^2P\sqrt{T}$	$\dot{W}_f/\delta\sqrt{\theta}$

$\theta = T/T_{std} = T/519$ ($T/288$) = corrected temperature
$\delta = P/p_{std} = P/14.7$ ($P/1.013 \times 10^5$) = corrected pressure
$\dot{W}_f$ = fuel flow

8.2.2 Turbofan

Most large airliners and high subsonic transport aircraft are powered by turbofan engines. Typical commercial engine thrust ratings range from 7000 lb (31 kN) to 90 000 lb (400 kN+) suitable for large aircraft such as the Boeing 747. The turbofan is characterized by an oversized fan compressor stage at the front of the engine which bypasses most of the air around the outside of the engine where it rejoins the exhaust gases at the back, increasing significantly the available thrust. A typical bypass ratio is 5–6 to 1. Turbofans have better efficiency than simple turbojets because it is more efficient to accelerate a large mass of air moderately through the fan to develop thrust than to highly accelerate a smaller mass of air through the core of the engine (i.e. to develop the same thrust). Figure 8.3 shows the basic turbofan and Figure 8.6 its two- and three-spool variants. The two-spool arrangement is the most common, with a single stage fan plus turbine

High pressure (hp) spool: The hp turbine (HPT)drives the high pressure compressor (HPC)

Two spool (most common aero-engine configuration)

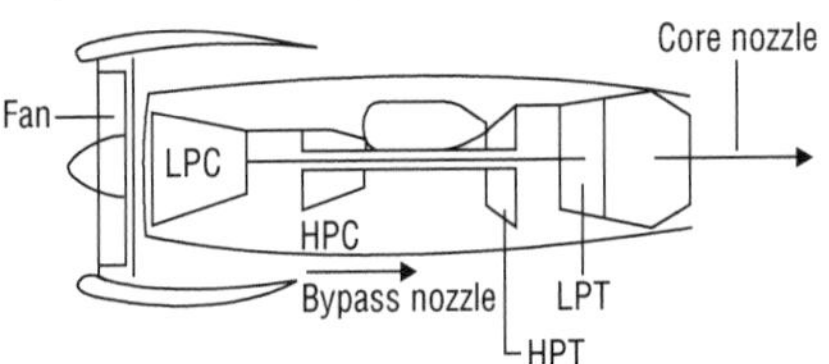

Low pressure spool: the lp turbine (LPT) drives the low pressure compressor (LPC)

Three spool engine (Rolls-Royce RB211)

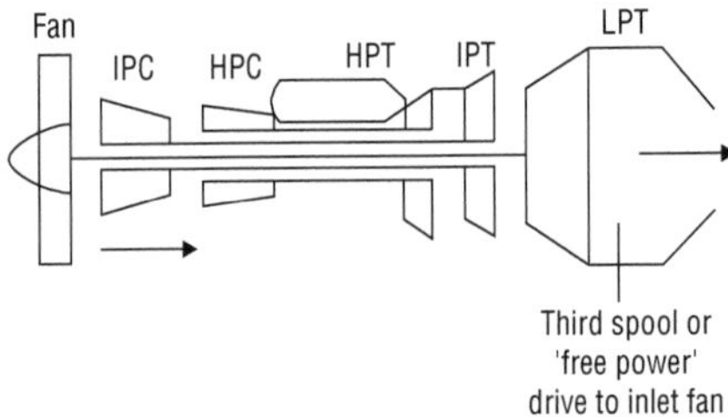

Fig. 8.6 Turbofan: 2- and 3-spool variants

on the low pressure rotor and an axial compressor plus turbine on the high pressure rotor. Many turbines are fitted with thrust reversing cowls that act to reverse the direction of the slipstream of the fan bypass air.

8.2.3 Turboprop

The turboprop configuration is typically used for smaller aircraft. Data for commercial models are shown in Table 8.2. The engine (see Figure 8.3) uses a separate power turbine stage to provide torque to a forward-mounted propeller. The propeller thrust is augmented by gas thrust from the exhaust. Although often overshadowed by the turbofan, recent developments in propeller technology mean that smaller airliners such as the SAAB 2000 (2 × 4152 hp (3096 kW) turboprops) can compete on speed and fuel cost with comparably sized turbofan aircraft. The most common turboprop configuration is a single shaft with centrifugal compressor and integral gearbox. Commuter airliners often use a two- or three-shaft ‘free turbine’ layout.

8.2.4 Propfans

Propfans are a modern engine arrangement specifically designed to achieve low fuel consumption. They are sometimes referred to as inducted fan engines. The most common arrangement is a two-spool gas generator and aft-located gearbox driving a ‘pusher’ fan. Historically, low fuel prices have reduced the drive to develop propfans as commercially viable mainstream engines. Some Russian aircraft such as the Anotov An-70 transport have been designed with propfans.

8.2.5 Turboshafts

Turboshaft engines are used predominantly for helicopters. A typical example such as the Rolls-Royce Turbomeca RTM 32201 has a three-stage axial compressor direct-coupled to a two-stage compressor turbine, and a two-stage

Table 8.2 Aircraft engines – basic data

Company	*Allied Signal*	*CFE*	*CFMI*	*General Electric (GE)*			*IAE (PW, RR, MTU, JAE)*		*Pratt & Witney*				*Rolls-Royce*			*ZMKB*
Engine type/Model	LF507	CFE738	CFM 56 5C2	CF34 3A,3B	CF6 80E1A2	GE 90 85B	V2522 A5	V2533 A5	PW4052	PW4056	PW4168	PW4084	TRENT 772	TAY 611	RB-211-524H	D-436T1
Aircraft	BA146-300 Avro RJ	Falcon 2000	A340	Canadair RJ	A330	B777-200/300	MD90-10/30 A319	A321-200	B767-200 &200ER	B747-400 767-300ER	A330	B777	A330	F100.70 Gulfst V	B747-400 B767-300	Tu-334-1 An 72,74
In service date	1991	1992	1994	1996	1995		1993	1994	1986	1987	1993	1994	1995	1988	1989	1996
Thrust (lb)	7000	5918	31 200	9220	67 500	90 000	22 000	33 000	52 200	56 750	68 000	84 000	71 100	13 850	60 600	16 865
Flat rating (°C)	23	30	30		30	30	30	30	33.3	33.3	30	30	30	30	30	30
Bypass ratio	5.6	5.3	6.4				5	4.6	4.85	4.85	5.1	6.41	4.89	3.04	4.3	4.95
Pressure ratio	13.8	23	31.5	21	32.4	39.3	24.9	33.4	27.5	29.7	32	34.2	36.84	15.8	33	25.2
Mass flow (lb/s)	256	240	1065		1926	3037	738	848	1705	1705	1934	2550	1978	410	1605	
SFC (lb/hr/lb)	0.406	0.369	0.32	0.35	0.33		0.34	0.37	0.351	0.359				0.43	0.563	
Climb																
Max thrust (lb)			7580			18 000	5550	6225					15 386	3400	12 726	
Flat rating (°C)							ISA+10	ISA+10					ISA+10	ISA+5	ISA+10	

Cruise																
Altitude (ft)		40 000	35 000			35 000	35 000	35 000	35 000	35 000	35 000	35 000	35 000	35 000	35 000	36 089
Mach number		0.8	0.8			0.83	0.8	0.8	0.8	0.8	0.8	0.83	0.82	0.8	0.85	0.75
Thrust (lb)		1310					5185	5725					11500	2550	11813	3307
Thrust lapse rate							0.2	0.174					0.162	0.184	0.195	0.196
Flat rating (°C)							ISA+10	ISA+10					ISA+10		ISA+10	
SFC (lb/hr/lb)	0.414	0.645	0.545		0.562	0.545	0.574	0.574					0.565	0.69	0.57	0.61
Dimensions																
Length (m)	1.62	2.514	2.616	2.616	4.343	5.181	3.204	3.204	3.879	3.879	4.143	4.869	3.912	2.59	3.175	
Fan diameter (m)	1.272	1.219	1.945	1.245	2.794	3.404	1.681	1.681	2.477	2.477	2.535	2.845	2.474	1.52	2.192	1.373
Basic eng. weight (lb)	1385	1325	5700	1670	10 726	16 644	5252	5230	9400	9400	14 350	13 700	10 550	2951	9670	3197
Layout																
Number of shafts	2	2	2	2	2	2	2	2	2	2	2	2	3	2	3	3
Compressor	various	1+5LP +1CF	1+4LP 9HP	1F +14cHP	1+4LP 14HP	1+3LP 10HP	1+4LP 10HP	1+4LP 10HP	1+4LP 11HP	1+4LP 11HP	1+5LP 11HP	1+6LP 11HP	1LP 8IP 6HP	1+3LP 12HP	1LP 7IP 6HP	1+1L 6I 7HP
Turbine	2HP 2LP	2HP 3LP	1HP 5LP	2HP 4LP	2HP 5LP	2HP 6LP	2HP 5LP	2HP 5LP	2HP 4LP	2HP 4LP	2HP 5LP	2HP 7LP	1HP 1IP 4LP	2HP 3LP	1HP 1IP 3LP	1HP 1IP 3LP

power turbine. Drive is taken off the power turbine shaft, through a gearbox, to drive the main and tail rotor blades. Figure 8.3 shows the principle.

8.2.6 Ramjet

This is the crudest form of jet engine. Instead of using a compressor it uses 'ram effect' obtained from its forward velocity to accelerate and pressurize the air before combustion. Hence, the ramjet must be accelerated to speed by another form of engine before it will start to work. Ramjet-propelled missiles, for example, are released from moving aircraft or accelerated to speed by booster rockets. A supersonic version is the *scramjet* which operates on liquid hydrogen fuel.

8.2.7 PULSEJET

A pulsejet is a ramjet with an air inlet which is provided with a set of shutters fixed to remain in the closed position. After the pulsejet engine is launched, ram air pressure forces the shutters to open, and fuel is injected into the combustion chamber and burned. As soon as the pressure in the combustion chamber equals the ram air pressure, the shutters close. The gases produced by combustion are forced out of the jet nozzle by the pressure that has built up within the combustion chamber. When the pressure in the combustion chamber falls off, the shutters open again, admitting more air, and the cycle repeats.

8.3 Engine data lists

Table 8.2 shows indicative design data for commercially available aero engines from various manufacturers.

8.4 Aero engine terminology

See Table 8.3.

Table 8.3

Afterburner A tailpipe structure attached to the back of military fighter aircraft engine which provides up to 50% extra power for short bursts of speed. Spray bars in the afterburner inject large quantities of fuel into the engine's exhaust stream.
Airflow Mass (weight) of air moved through an engine per second. Greater airflow gives greater thrust.
Auxiliary power Units (APUs) A small (< 450 kW) gas turbine used to provide ground support power.
Bleed air Air taken from the compressor section of an engine for cooling and other purposes.
Bypass Ratio (BPR) The ratio of air ducted around the core of a turbofan engine to the air that passes through the core. The air that passes through the core is called the primary airflow. The air that bypasses the core is called the secondary airflow. Bypass ratio is the ratio between secondary and primary airflow.
Combustion chamber The section of the engine in which the air passing out of the compressor is mixed with fuel.
Compressor The sets of spinning blades that compress the engine air stream before it enters the combustor. The air is forced into a smaller and smaller area as it passes through the compressor stages, thus raising the pressure ratio.
Compressor Pressure Ratio (CPR) The ratio of the air pressure exiting the compressor compared to that entering. It is a measure of the amount of compression the air experiences as it passes through the compressor stage.
Core engine A term used to refer to the basic parts of an engine including the compressor, diffuser/combustion chamber and turbine parts.
Cowl The removable metal covering of an aero engine.
Diffuser The structure immediately behind an engine's compressor and immediately in front of the combustor. It slows down compressor discharge air and prepares the air to enter the combustion chamber at a lower velocity so that it can mix with the fuel properly for efficient combustion.

Table 8.3 ***Continued***

Digital Electronic Engine Control (DEEC)
The computer that automatically controls all the subsystems of the engine.

Electronic Engine Control (EEC)
Also known as the FADEC (full-authority digital electronic engine control), it is an advanced computer which controls engine functions.

Engine Build Unit (EBU)
The equipment supplied by the aircraft manufacturer that is attached to the basic engine, e.g. ducting, wiring packages, electrical and hydraulic pumps and mounting parts.

Engine Pressure Ratio (EPR)
The ratio of the pressure of the engine air at the rear of the turbine section compared to the pressure of the air entering the compressor.

Exhaust Gas Temperature (EGT)
The temperature of the engine's gas stream at the rear of the turbine stages.

Fan
The large disc of blades at the front of a turbofan engine.

In-flight Shutdown Rate (IFSD)
A measure of the reliability of an engine, expressed as the number of times per thousand flight hours an engine must be shut down in flight.

Inlet duct
The large round structure at the front of an engine where the air enters.

Line Replaceable Unit (LRU)
An engine component that can be replaced 'in service' at an airport.

Mean Time Between Failures (MTBF)
The time that a part or component operates without failure.

Nacelle
The cylindrical structure that surrounds an engine on an aircraft. It contains the engine and thrust reverser and other mechanical components that operate the aircraft systems.

N1 (rpm)
The rotational speed of the engine's low pressure compressor and low pressure turbine stage.

N2 (rpm)
The rotational speed of the engine's high pressure compressor.

Table 8.3 ***Continued***

Nozzle
The rear portion of a jet engine in which the gases produced in the combustor are accelerated to high velocities.

Pressure ratio
The ratio of pressure across the compression stage (or turbine stages) of an engine.

A surge
A disturbance of the airflow through the engine's compressor, often causing 'stall' of the compressor blades

Thrust
A measurement of engine power.

Thrust reverser
A mechanical device that redirects the engine exhaust and air stream forward to act as a brake when an aircraft lands. The rotating parts of the engine do not change direction; only the direction of the exhaust gases.

Thrust specific fuel consumption
The mass (weight) of fuel used per hour for each unit of thrust an engine produces.

Turbine
The turbine consists of one or more rows of blades mounted on a disc or drum immediately behind the combustor. Like the compressor, the turbine is divided into a low pressure and a high pressure section. The high pressure turbine is closest to the combustor and drives the high pressure compressor through a shaft connecting the two. The low pressure turbine is next to the exhaust nozzle and drives the low pressure compressor and fan through a separate shaft.

8.5 Power ratings

Figure 8.7 shows comparative power ratings for various generic types of civil and military aircraft.

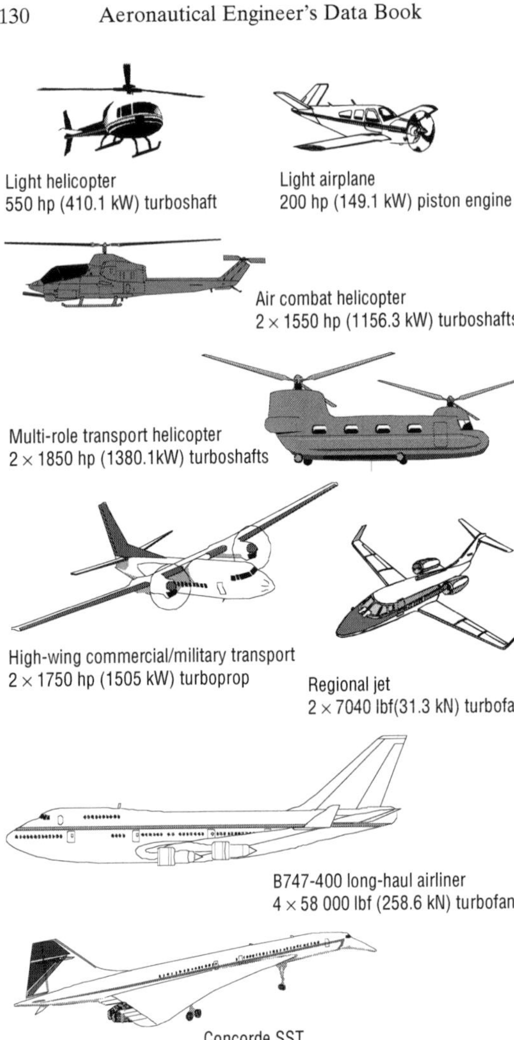
Light helicopter
550 hp (410.1 kW) turboshaft
Light airplane
200 hp (149.1 kW) piston engine
Air combat helicopter
2 × 1550 hp (1156.3 kW) turboshafts
Multi-role transport helicopter
2 × 1850 hp (1380.1kW) turboshafts
High-wing commercial/military transport
2 × 1750 hp (1505 kW) turboprop
Regional jet
2 × 7040 lbf(31.3 kN) turbofan
B747-400 long-haul airliner
4 × 58 000 lbf (258.6 kN) turbofan
Concorde SST
4 × 38 000 lbf (169.4 kN) turbojet with reheat
B777-300 airliner
2 × 84 700 lbf (377 kN) turbofan

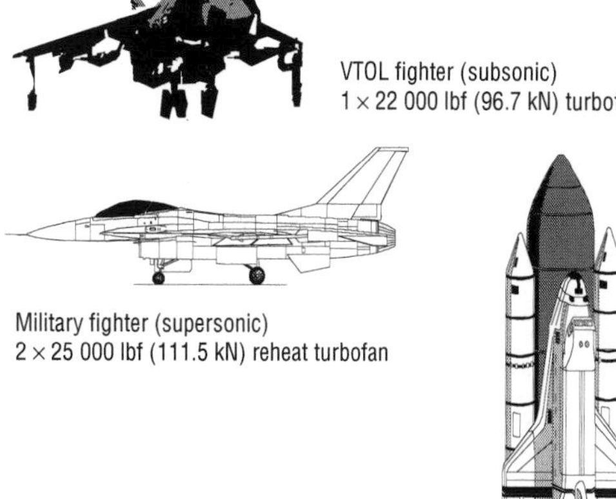

Fig. 8.7 Aircraft comparative power outputs

Section 9

Aircraft performance

9.1 Aircraft roles and operational profile

Civil aircraft tend to be classified mainly by range. The way in which a civil aircraft operates is termed its *operational profile*. In the military field a more commonly used term is *mission profile*. Figure 9.1 shows a typical example and Table 9.1 some commonly used terms.

9.1.1 Relevant formula

Relevant formulae used during the various stages of the operational profile are:

Take-off ground roll

$$S_G = 1/(2gK_A).\ln[K_T + K_A.V^2_{LOF})/K_T].$$

This is derived from $\int_0^{V_{LOF}} [(\frac{1}{2}a)dV^2]$

Total take-off distance

$$S_{TO} = (S_G)(F_{p1})$$

where F_{p1} is a 'take-off' plane form coefficient between about 1.1 and 1.4.

$$V_{TRANS} = (V_{LOF} + V_2)/2 \cong 1.15V_S$$

Rate of climb

For small angles, the rate of climb (RC) can be determined from:

$$RC = \frac{(F - D)V}{W\left(1 + \frac{V}{g}\frac{dV}{dh}\right)}$$

where $V/g.\ dV/dh$ is the correction term for flight acceleration

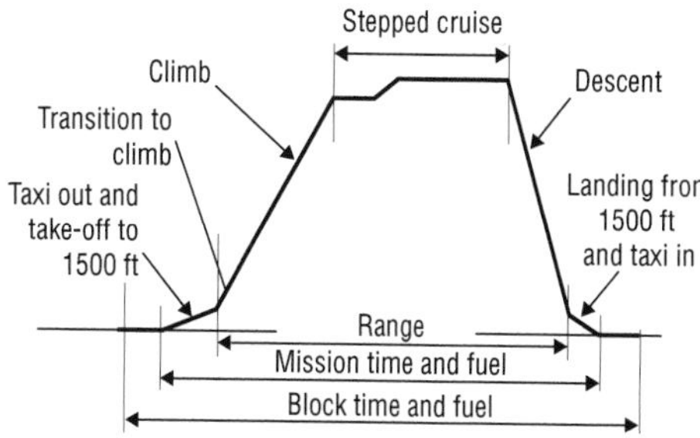

Fig. 9.1 A typical operational profile

Table 9.1 Operational profile terms

Take off	Take-off run available: operational length of the runway. Take-off distance available: length of runway including stopway (clear area at the end) and clearway (distance from end of stopway to the nearest 35 ft high obstruction). V_s: aircraft stall speed in take-off configuration. V_R: rotate speed. V_2: take-off climb speed at 35 ft clearance height. V_{mc}: minimum speed for safe control. V_{LOF}: Lift off speed: speed as aircraft clears the ground.
Transition to climb	V_{TRANS}: average speed during the acceleration from V_{LOF} to V_2. γ: final climb gradient.
Take-off climb	γ_c: best climb angle. 1st segment: first part of climb with undercarriage still down. 2nd segment: part of climb between 'undercarriage up' and a height above ground of 400 ft. 3rd segment: part of climb between 400 ft and 1500 ft.
Climb from 1500 ft to cruise	1st segment: part of climb between 1500 ft and 10 000 ft. 2nd segment: part of climb from 10 000 ft to initial cruise altitude. V_c: rate of climb.
Cruise	V_T: cruise speed.
Descent	V_{mc}: speed between cruise and 10 000 ft. (See Figure 9.2 for further details.)
Landing	Approach: from 50 ft height to flare height (h_f). Flare: deceleration from approach speed (V_A) to touch down speed V_{TD}. Ground roll: comprising the free roll (no brakes) and the braked roll to a standstill.

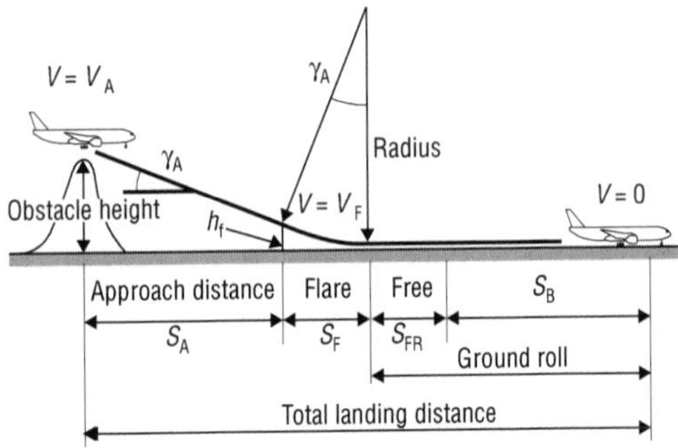

Fig. 9.2 Approach and landing definitions

F = thrust
g = acceleration due to gravity
h = altitude
RC = rate of climb
S = reference wing area
V = velocity
W = weight
W_f = fuel flow

Flight-path gradient

$$\gamma = \sin^{-1}\left(\frac{F - D}{W}\right)$$

Time to climb

$$\Delta t = \frac{2(h_2 - h_1)}{(\text{RC})_1 + (\text{RC})_2}$$

Distance to climb

$$\Delta S = V(\Delta t)$$

Fuel to climb

$$\Delta\text{Fuel} = W_f(\Delta t)$$

Cruise

The basic cruise distance can be determined by using the Breguet range equation for jet aircraft, as follows:

Cruise range

$$R = L/D(V/\text{sfc})\ln(W_0/W_1)$$

where subscripts '0' and '1' stand for initial and final weight, respectively.

Cruise fuel

$$\text{Fuel} = W_0 - W_1 = W_f(e^{R/k-1})$$

where k, the range constant, equals $L/D(V/\text{sfc})$ and R = range.

Cruise speeds
Cruise speed schedules for subsonic flight can be determined by the following expressions.

Optimum mach number (M_{DD}), optimum-altitude cruise
First calculate the atmospheric pressure at altitude:

$$P = \frac{W}{0.7(\mathrm{M}_{\mathrm{DD}}^2)(C_{L_{\mathrm{DD}}})S}$$

where M_{DD}^2 = drag divergence Mach number.

Then input the value from cruise-altitude determination graph for cruise altitude.

Optimum mach number, constant-altitude cruise
Optimum occurs at maximum $\mathrm{M}(L/D)$.

$$\mathrm{M} = \sqrt{\frac{W/S}{0.7P}\sqrt{\frac{3K}{C_{D_{\min}}}}}$$

where K = parabolic drag polar factor
P = atmospheric pressure at altitude

Landing
Landing distance calculations cover distance from obstacle height to touchdown and ground roll from touchdown to a complete stop.

Approach distance

$$S_{air} = \left(\frac{V^2_{obs} - V^2_{TD}}{2g} + h_{obs}\right)(L/D)$$

where V_{obs} = speed at obstacle, V_{TD} = speed at touchdown, h_{obs} = obstacle height, and L/D = lift-to-drag ratio.

Landing ground roll

$$S_{gnd} = \frac{(W/S)}{g\rho(C_D - \mu_{BRK}C_L)} \ln\left[1 - \frac{A^2\,(C_D - \mu_{BRK}C_L}{((F/W) - \mu_{BRK}C_{Lm\times s})}\right]$$

9.2 Aircraft range and endurance

The main parameter is the *safe operating range*; the furthest distance between airfields that an aircraft can fly with sufficient fuel allowance for headwinds, airport stacking and possible diversions. A lesser used parameter is the *gross still air range*; a theoretical range at cruising height between airfields. Calculations of range are complicated by the fact that total aircraft mass decreases as a flight progresses, as the fuel mass is burnt (see Figure 9.3). *Specific air range* (r) is defined as distance/fuel used (in a short time). The equivalent endurance term is *specific endurance* (e).

General expressions for range and endurance can be shown to follow the models in Table 9.2.

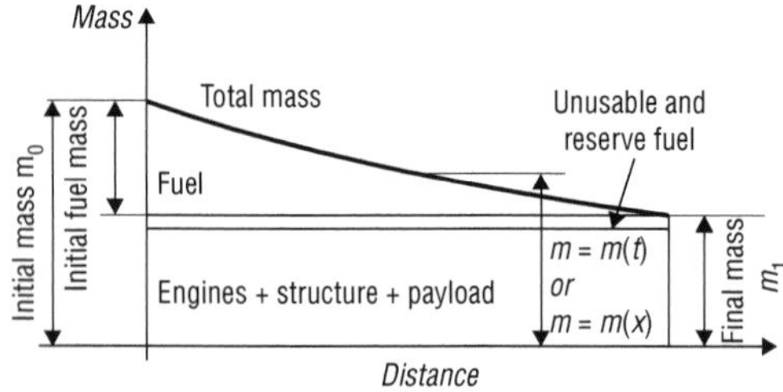

Fig. 9.3 Range terminology

Table 9.2 Range and endurance equations

	Propeller aircraft	*Jet aircraft*
Specific range (r)	$r = \eta/fD$	$r = V/f_jD$
Specific endurance (e)	$e = \eta/fDV$	$e = 1/f_jD$
Range (R)	$R = \int_{m_1}^{m_0} \frac{\eta dm}{fD} = \int_{m_1}^{m_0} \frac{\eta}{f}\left(\frac{C_L}{C_D}\right)\frac{dm}{mg}$	$R = \int_{m_1}^{m_0} \frac{V dm}{f_jD} = \int_{m_1}^{m_0} \frac{V}{f_j}\left(\frac{C_L}{C_D}\right)\frac{dm}{mg}$
Endurance (E)	$E = \int_{m_1}^{m_0} \frac{\eta dm}{fDV} = \int_{m_1}^{m_0} \frac{\eta}{fV}\left(\frac{C_L}{C_D}\right)\frac{dm}{mg}$	$E = \int_{m_1}^{m_0} \frac{dm}{f_jD} = \int_{m_1}^{m_0} \frac{1}{f_j}\left(\frac{C_L}{C_D}\right)\frac{dm}{mg}$

9.3 Aircraft design studies

Aircraft design studies are a detailed and iterative procedure involving a variety of theoretical and empirical equations and complex parametric studies. Although aircraft specifications are built around the basic requirements of payload, range and performance, the design process also involves meeting overall criteria on, for example, operating cost and take-off weights.

The problems come from the *interdependency* of all the variables involved. In particular, the dependency relationships between wing area, engine thrust and take-off weight are so complex that it is often necessary to start by looking at existing aircraft designs, to get a first impression of the practicality of a proposed design. A design study can be thought of as consisting of two parts: the initial 'first approximations' methodology, followed by 'parametric estimate' stages. In practice, the processes are more iterative than purely sequential. Table 9.3 shows the basic steps for the initial 'first approximations' methodology, along with some general *rules of thumb*.

Figure 9.4 shows the basis of the following stage, in which the results of the initial estimates are used as a basis for three alternatives for wing area. The process is then repeated by estimating three values for take-off

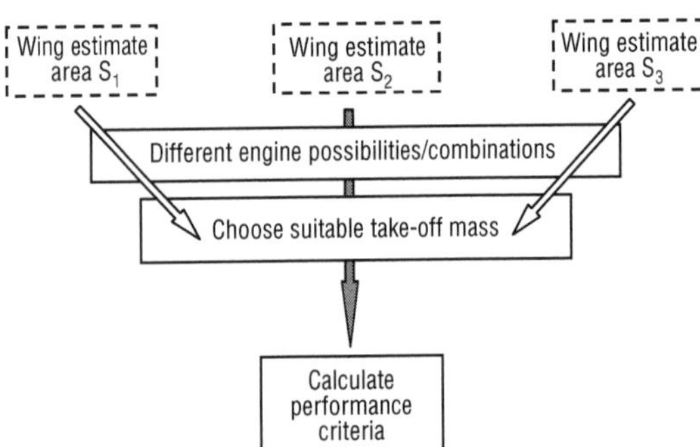

Fig. 9.4 A typical 'parametric' estimate stage

Table 9.3 The 'first approximations' methodology

Estimated parameter	*Basic relationships*	*Some 'rules of thumb'*
1. Estimate the wing loading W/S.	$W/S = 0.5\,\rho V^2 C_L$ in the 'approach' condition.	Approach speed lies between 1.45 and 1.62 V_{stall}. Approach C_L lies between $C_{L\max}/2.04$ and $C_{L\max}/2.72$.
2. Check C_L in the cruise.	$C_L = \dfrac{0.98(W/S)}{q}$ where $q = 0.5\,\rho V^2$	C_L generally lies between 0.44 and 0.5.
3. Check gust response at cruise speed.	Gust response parameter $= \dfrac{\alpha_{1wb}.\mathrm{AR}}{(W/S)}$ α_{1wb} is the wing body lift curve slope obtained from data sheets.	
4. Estimate size.	Must comply with take-off and climb performance.	Long range aircraft engines are sized to 'top of climb' requirements.
5. Estimate take-off wing loading and T/W ratio as a function of C_{LV2}	$s = kM^2g^2/(S_w T.C_{LV2}\,\sigma)$	$1.7 < C_{L\max} < 2.2$ $1.18 < C_{LV_2} < 1.53$
6. Check the capability to climb (gust control) at initial cruise altitude.	Cruise L/D is estimated by comparisons with existing aircraft data. $F_n/M_{CL} = (L/D)^{-1} + (300/101.3V)$ (imperial units)	$17 < L/D < 21$ in the cruise for most civil airliners.
7. Estimate take-off mass	$M_{TO} = M_E + M_{PAY} + M_f$	$0.46 < \dfrac{\mathrm{OEM}}{\mathrm{MTOM}} < 0.57$

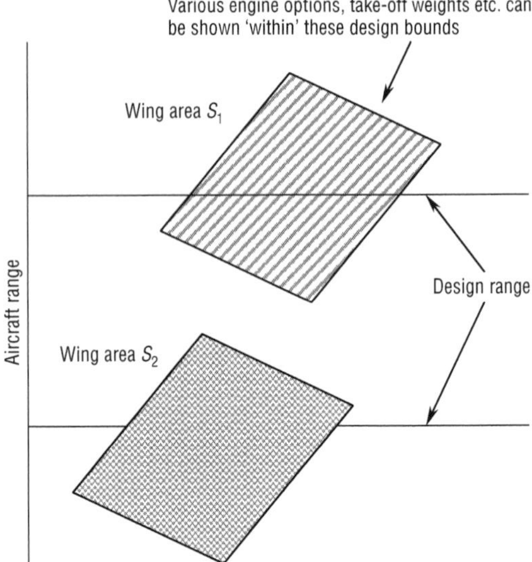

Fig. 9.5 Typical parametric plot showing design 'bounds'

weight and engine size for each of the three wing area 'conclusions'. The results are then plotted as parametric study plots and graphs showing the bounds of the various designs that fit the criteria chosen (Figure 9.5).

9.3.1 Cost estimates

Airlines use their own (often very different) standardized methods of estimating the capital and operating cost of aircraft designs. They are complex enough to need computer models and all suffer from the problems of future uncertainty.

9.4 Aircraft noise

Airport noise levels are influenced by FAR-36 which sets maximum allowable noise levels for subsonic aircraft at three standardized measurement positions (see Figure 9.6). The maximum allowable levels set by FAR-36 vary, depending on aircraft take-off weight (kg).

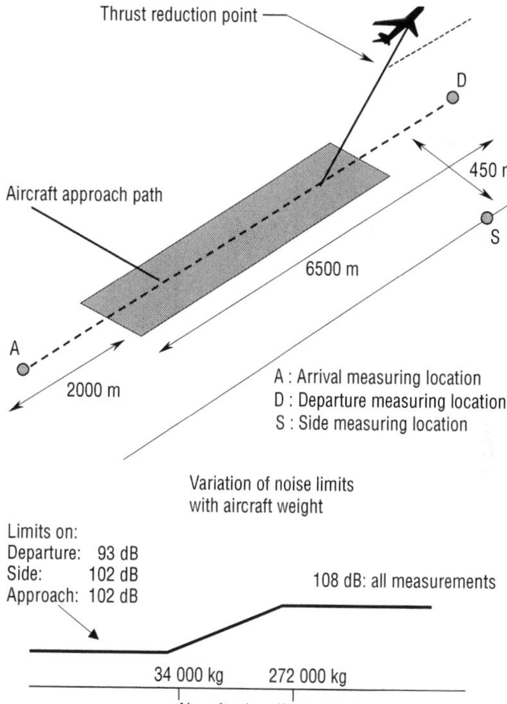

Fig. 9.6 Airport noise measurement locations

9.4.1 Aircraft noise spectrum

The nature of an aircraft's noise spectrum and *footprint* depends heavily on the type of engine used. Some rules of thumb are:

- The predominant noise at take-off comes from the aircraft engines.
- During landing, 'aerodynamic noise' (from pressure changes around the airframe and control surfaces) becomes more significant, as the engines are operating on reduced throttle settings.
- Low bypass ratio turbofan engines are generally noisier than those with high bypass ratios.
- Engine noise energy is approximately proportional to (exhaust velocity)^7.

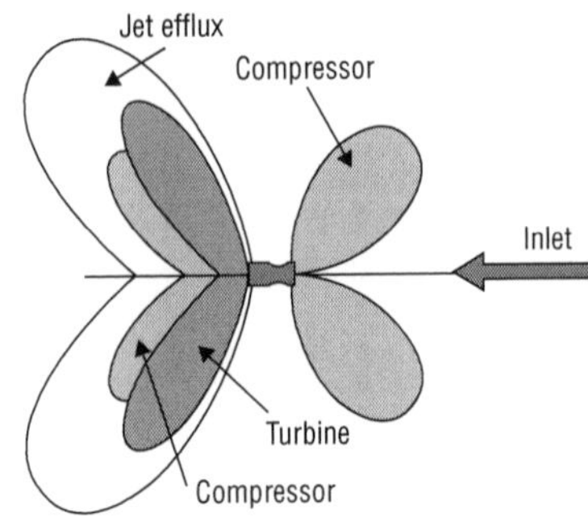

The general aircraft noise 'footprint'

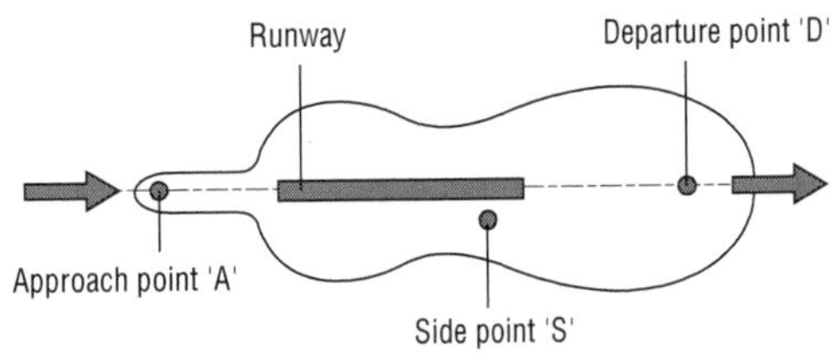

Noise footprint shape for four-engine passenger jet

Fig. 9.7 Aircraft noise characteristics

Figure 9.7 shows the general shape of an aircraft noise footprint and the resulting distribution of noise in relation to the runway and standardized noise measurement points.

Supersonic aircraft such as Concorde using pure turbojet engines require specific noise reduction measures designed to minimize the noise level produced by the jet efflux. Even using 'thrust cutback' and all possible technical developments, supersonic aircraft are still subject to severe restrictions in and around most civil aviation airports.

Sonic booms caused by low supersonic Mach numbers (< MA 1.15) are often not heard at ground level, as they tend to be refracted upwards. In some cases a portion of

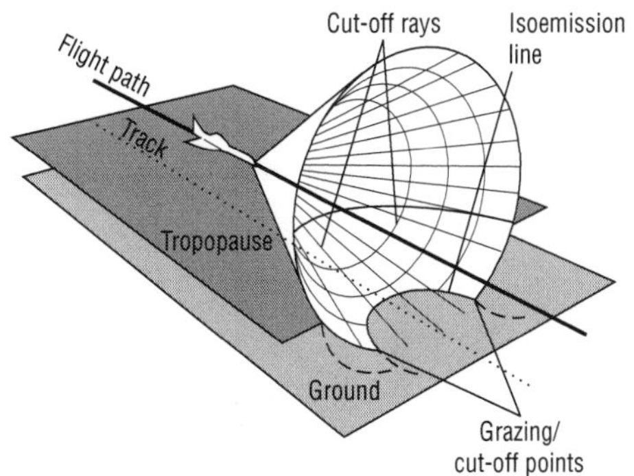

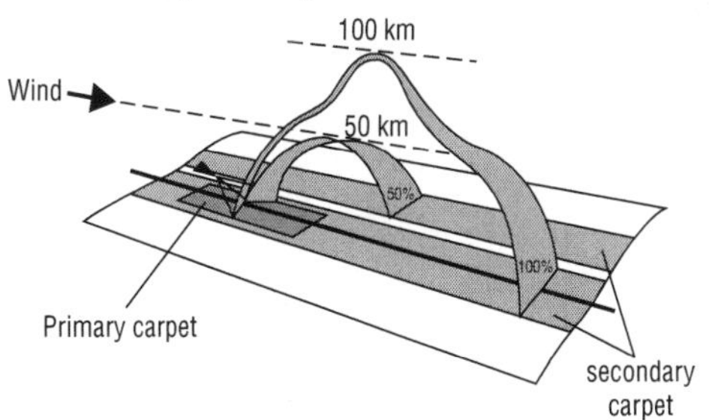

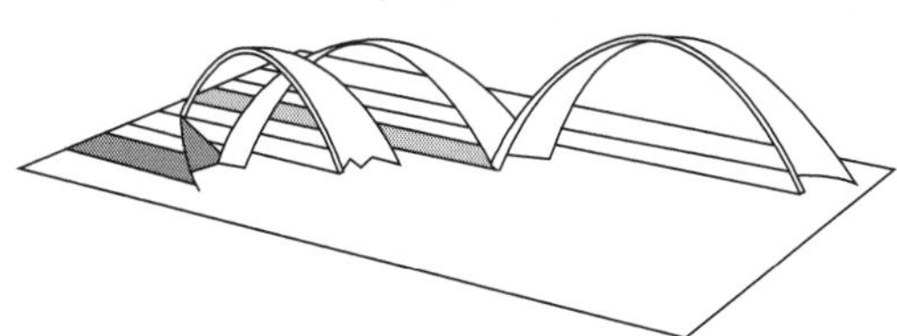

'Bouncing' shock waves giving refracted and reflected booms at greatly reduced sound pressure

Fig. 9.8 Sonic boom characteristics

the upward-heading wave may be refracted back to the surface, forming a 'secondary boom' at greatly reduced sound pressure. Shock waves may also bounce, producing sound levels only slightly above ambient noise level (see Figure 9.8)

9.5 Aircraft emissions

Aircraft engine emissions vary with the type of engine, the fuel source used, and the operational profile. Emission levels are governed by ICAO recommendations. For comparison purposes the flight profile is divided into the take-off/landing segment and the cruise segment (designated for these purposes as part of the flight profile above 3000 ft). Table 9.4 shows an indicative 'emission profile' for a large four-engined civil aircraft.

Table 9.4 An indicative 'emission profile'

	Emissions in g/kg fuel			
	CO	*NO_x**	*SO_2*	*HC (unburnt)*
Take-off	0.4	27	0.5	0.06
Cruise >3000 ft*	No agreed measurement method. Varies with aircraft and flight profile			
Approach/landing	2.0	11	0.5	0.12

*Some authorities use a NO_x emission index as a general measure of the level of 'amount of pollution' caused per unit of fuel burnt.

Section 10

Aircraft design and construction

10.1 Basic design configuration

Basic variants for civil and military aircraft are shown in Figure 10.1 Large civil airliners have a low wing design in which the wing structure passes through the freight area beneath the passenger cabin. Small airliners may use the high wing design, with a bulge over the top line of the fuselage so as not to restrict passenger headroom. Having a continuous upper surface to the wing (as in the high-wing layout) can improve the *L*/*D* ratio and keeps the engines at a higher distance from the ground, so avoiding debris from poor or unpaved runways.

Tailplane configuration is matched to the wing type and includes high tail, low tail, flat, vee and dihedral types. Low tails increase stability at high angles of attack but can also result in buffeting (as the tail operates in the wing wake) and non-linear control response during normal flight. High tails are generally necessary with rear-fuselage mounted engines and are restricted to high speed military aircraft use. Figure 10.2 shows variants in tail and engine position. The rear-engine configuration has generally been superseded by under-wing mounted engines which optimizes bending moments and enables the engine thrust loads to be fed directly into the wing spars. In contrast, rear-fuselage mounted engines decrease cabin noise.

10.1.1 Aspect ratio (AR)

The aspect ratio (AR) is a measure of wingspan in relation to mean wing chord. Values for subsonic aircraft vary between about 8 and 10 (see Tables 10.1 and 10.2). Figure 10.1 shows some typical configurations.

Low wing

High wing

Straight-wing turboprop

AR=10.5

High-wing turbofan

AR=8.9

Twin engine Airbus

AR=9.4

Concorde

AR=2.1

Flying wing

Four engine military bomber

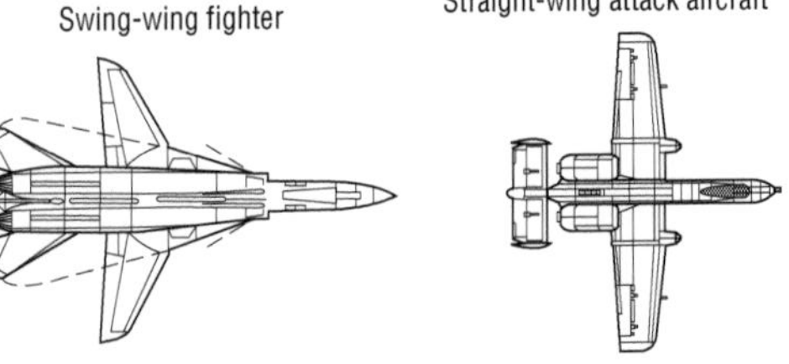

Fig. 10.1 Basic design configurations

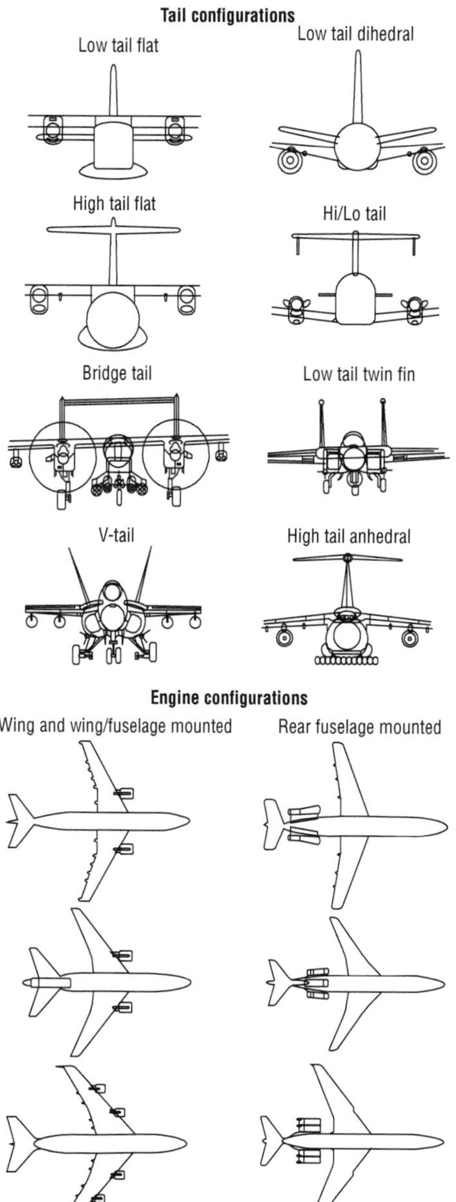

Fig. 10.2 Variants in tail and engine position

Table 10.1 Civil aircraft – basic data

Manufacturer *Type* *Model*	*Airbus* *A320–* *200*	*Airbus* *A321–* *200*	*Airbus* *A330–* *200*	*Airbus* *A340–* *300*	*Airbus* *A340–* *500*	*Boeing* *717–* *200*	*Boeing* *737–* *800*	*Cadair* *Reg. Jet* *100ER*	*Embraer* *EMB-145*	*Fokker* *F70*	*Fokker* *F100*	*Ilyushin* *Il-96M*	*McDon.* */Doug.* *MD-90-30*	*McDon.* */Doug.* *MD-11*	*Tupolev* *Tu-204* *-200*
Initial service date	1988	1993	1998	1994	2002	1999	1998	1992	1997	1988	1988	1996	1995	1990	1997
Engine manufacturer	CFMI	CFMI	GE	CFMI	R-R	BMW R-R	CFMI	GE	Allison	R-R	R-R		IAE	GE	Soloviev
Model/Type	CFM56-5A3	CFM56-5B3	CF6-80E1A4	CFM-56-5C4	Trent 553	715	CFM56-7B24	CF34-3A1	AE3007A	Tay 620	Tay 620	2337	V2525-D5	CF6-80 C2 DIF	PS-90A
No. of engines	2	2	2	4	4	2	2	2	2	2	2	4	2	3	2
Static thrust (kN)	111.2	142	310	151	235.8	97.9	107	41	31.3	61.6	61.6	164.6	111.2	274	157
Accommodation:															
Max. seats (single class)	179	220	380	440	440	110	189	52	50	79	119	375	182	405	214
Two class seating	150	186	293	335	350	106	160	–	–	70	107	335	153	323	196
Three class seating	–	–	253	295	313	–	–	–	–	–	–	312		293	190
No. abreast	6	6	9	9	9	5	6	4	3	–	–	9	5	10	6
Hold volume (m^3)	38.76	51.76	136	162.9	134.1	25	47.1	14.04	13.61	12.78	16.72	143.04	38.03	194	26.4
Volume per passenger	0.22	0.24	0.36	0.37	0.3	0.23	0.25	0.27	0.27	0.16	0.14	0.38	0.21	0.48	0.12
Mass (weight) (kg):															
Ramp	73 900	89 400	230 900	271 900	365 900	52 110	78 460	23 246	19 300	36 965	43 320		71 215	285 081	111 750
Max. take-off	73 500	89 000	230 000	271 000	365 000	51 710	78 220	23 133	19 200	36 740	43 090	270 000	70 760	283 720	110 750
Max. landing	64 500	73 500	177 150	190 000	236 000	46 266	65 310	21 319	18 700	34 020	38 780	175 158	64 410	207 744	89 500

Zero-fuel	60 500	71 500	165 142	178 000	222 000	43 545	61 680	19 958	17 100	31 975	35 830	190 423	58 965	195 043	84 200
Max. payload	19 190	22 780	36 400	48 150	51 635	12 220	14 690	6295	5515	9302	11 108	58 000	17 350	55 566	25 200
Max. fuel payload	13 500	19 060	–	33 160	31 450	8921	15 921	3006	3498	6355	7805	17 290	13 659	30 343	18 999
Design payload	14 250	17 670	24 035	28 025	29 735	10 070	15 200	4940	4750	6650	10 165	29 640	14 535	30 685	18 620
Design fuel load	17 940	23 330	85 765	113 125	164 875	9965	21 540	4530	2865	7417	8332	107 960	16 810	118 954	33 130
Operational empty	41 310	48 000	120 200	129 850	170 390	31 675	41 480	13 663	11 585	22 673	24 593	132 400	39 415	134 081	59 000
Weight ratios:															
Ops empty/Max. T/O	0.562	0.539	0.523	0.479	0.467	0.613	0.53	0.591	0.603	0.617	0.571	0.49	0.557	0.473	0.533
Max. payload/Max. T/O	0.261	0.256	0.158	0.178	0.141	0.236	0.188	0.272	0.287	0.253	0.258	0.215	0.245	0.196	0.228
Max. fuel/Max. T/O	0.256	0.21	0.478	0.412	0.423	0.212	0.263	0.276	0.212	0.207	0.245	0.44	0.247	0.424	0.292
Max. landing/Max. T/O	0.878	0.826	0.77	0.701	0.647	0.895	0.835	0.922	0.974	0.926	0.9	0.649	0.91	0.732	0.808
Fuel (litres):															
Standard	23 860	23 700	139 090	141 500	195 620	13 892	26 024	8080	5146	9640	13 365	150 387	22 107	152 108	40 938
Dimensions fuselage:															
Length (m)	37.57	44.51	57.77	62.47	65.6	33	38.08	24.38	27.93	27.88	32.5	60.5	43	58.65	46.7
Height (m)	4.14	4.14	5.64	5.64	5.64	3.61	3.73					6.08	3.61	6.02	3.8
Width (m)	3.95	3.95	5.64	5.64	5.64	3.61	3.73					6.08	3.61	6.02	4.1
Finess ratio	9.51	11.27	10.24	11.08	11.63	4.3	7.4					9.95	11.91	9.74	11.39
Wing:															
Area (m^2)	122.4	122.4	363.1	363.1	437.3	92.97	124.6	54.54	51.18	93.5	93.5	391.6	112.3	338.9	182.4
Span (m)	33.91	33.91	58	58	61.2	28.4	34.3	20.52	20.04	28.08	28.08	55.57	32.87	51.77	40.3
MAC (m)	4.29	4.29	7.26	7.26	8.35	3.88	4.17	3.15	3.13	3.8	3.8	8.04	4.08	7.68	5.4
Aspect ratio	9.39	9.39	9.26	9.26	8.56	8.68	9.44	7.72	7.85	8.43	8.43	7.89	9.62	7.91	8.9
Taper ratio	0.24	0.24	0.251	0.251	0.22	0.196	0.278	0.288	0.231	0.235	0.235	0.279	0.195	0.239	0.228

Table 10.1 *Continued*

Manufacturer	*Airbus*	*Airbus*	*Airbus*	*Airbus*	*Airbus*	*Boeing*	*Boeing*	*Cadair*	*Embraer*	*Fokker*	*Fokker*	*Ilyushin*	*McDon. /Doug.*	*McDon. /Doug.*	*Tupolev*
Type	*A320–*	*A321–*	*A330–*	*A340–*	*A340–*	*717–*	*737–*	*Reg. Jet*							*Tu-204*
Model	*200*	*200*	*200*	*300*	*500*	*200*	*800*	*100ER*	*EMB-145*	*F70*	*F100*	*Il-96M*	*MD-90-30*	*MD-11*	*-200*
Average t/c %						11.6		10.83	11	10.28	10.28		11	9.35	
1/4 chord sweep (°)	25	25	29.7	29.7	31.1	24.5	25	24.75	22.73	17.45	17.45	30	24.5	35	28
High lift devices:															
Trailing edge flaps type	F1	F2	S2	S2	S2	S2	S2	S2	S2	F2	F2	S2	S2	S2	S2
Flap span/Wing span	0.78	0.78	0.665	0.665	0.625	0.65	0.599	0.66	0.72	0.58	0.58	0.79	0.63	0.7	0.77
Area (m^2)	21.1	21.1						10.6	8.36	17.08	17.08				
Leading edge flaps	slats	slats	slats	slats	slats	slats	slats/flaps	slats	none	none	none	slats	slats	slats	slats
Type															
Area (m^2)	12.64	12.64													
Vertical tail															
Area (m^2)	t21.5	21.5	47.65	45.2	47.65	19.5	23.13	9.18	7.2	12.3	12.3	56.2	21.4	56.2	34.2
Height (m)	6.26	6.26	9.44	8.45	9.44	4.35	6	2.6	3.1	3.3	3.3	8	4.7	11.16	7.7
Aspect ratio	1.82	1.82	1.87	1.58	1.87	0.97	1.56	0.74	1.33	0.89	0.89	1.14	1.03	2.22	1.73
Taper ratio	0.303	0.303	0.35	0.35	0.35	0.78	0.31	0.73	0.6	0.74	0.74	0.4	0.77	0.369	0.34
1/4 chord sweep (°)	34	34	45	45	45	45	35	41	32	41	41	45	43	40	36
Tail arm (m)	12.53	15.2	25.2	27.5	27.5	12.8	17.7	10.7	11.5	11.4	13.6	25.9	15.6	20.92	21.8
S_v/S	0.176	0.176	0.131	0.124	0.109	0.21	0.186	0.168	0.141	0.132	0.132	0.144	0.191	0.166	0.188
S_vL_v/S_b	0.065	0.079	0.057	0.059	0.049	0.095	0.096	0.088	0.081	0.053	0.064	0.067	0.09	0.067	0.101

Horizontal tail:															
Area (m^2)	31	31	31	72.9	93	24.2	32.4	9.44	11.2	21.72	21.72	96.5	33	85.5	44.6
Span (m)	12.45	12.45	12.45	19.06	21.5	10.8	13.4	6.35	7.6	10.04	10.04	20.57	12.24	18.03	15.1
Aspect ratio	5	5	5	4.98	4.97	4.82	5.54	4.27	5.16	4.64	4.64	4.38	4.54	3.8	5.11
Taper ratio	0.256	0.256	0.256	0.36	0.36	0.38	0.186	0.55	0.56	0.39	0.39	0.29	0.36	0.383	0.3
1/4 chord sweep (°)	29	29	29	30	30	30	30	30	17	26	26	37.5	30	35	34
Tail arm (m)	13.53	16.2	16.2	28.6	28.6	14.3	17.68	12.9	12.9	14.4	16	26.5	18.6	20.92	21.3
S_h/S	0.253	0.253	0.253	0.201	0.213	0.26	0.26	0.173	0.219	0.232	0.232	0.246	0.294	0.252	0.245
$S_h/L_h/S_c$	0.799	0.957	0.957	0.791	0.729	0.959	1.102	0.709	0.902	0.88	0.978	0.812	1.34	0.687	0.964
Undercarriage:															
Track (m)	7.6	7.6	7.6	10.7	10.7	4.88	5.7		4.1	5.04	5.04	10.4	5.09	10.6	7.82
Wheelbase (m)	12.63	16.9	16.9	25.4	28.53	17.6		11.39	14.45	11.54	14.01	27.35	23.53	24.6	17
Turning radius (m)	21.9	29	29	40.6				22.86		17.78	20.07			41	
No. of wheels (nose; main)	2;4	2;4	2;8	2;10	2;12	2;4	2;4	2;4	2;4	2;4	2;4	2;8	2;4	2;10	2;8
Main wheel diameter (m)	1.143	1.27					1.016	0.95	0.98	1.016	1.016	1.3			
Main wheel width (m)	0.406	0.455					0.368	0.3	0.31	0.356	0.356	0.48			
Nacelle:															
Length (m)	4.44	4.44	7	4.95	6.1	6.1	4.7	3.8	4	5.1	5.1	6	5.75	6.5	6
Max. width (m)	2.37	2.37	3.1	2.37	3.05	1.75	2.06	1.5	1.5	1.7	1.7	2.6	1.55	2.7	2.6
Performance Loadings:															
Max. power Load (kg/kN)	330.49	313.38	370.97	448.68	386.98	264.1	365.51	282.11	306.51	298.21	349.76	410.09	318.14	345.16	352.71
Max. wing Load (kg/m^2)	600.49	727.12	633.43	746.35	834.67	556.2	627.77	424.15	375.15	392.94	460.86	689.48	630.1	837.18	607.18
Thrust/Weight ratio	0.3084	0.3253	0.2748	0.2272	0.2634	0.386	0.2789	0.3613	0.3326	0.3418	0.2915	0.249	0.32	0.295	0.289

Table 10.1 *Continued*

Manufacturer *Type* *Model*	*Airbus* *A320–* *200*	*Airbus* *A321–* *200*	*Airbus* *A330–* *200*	*Airbus* *A340–* *300*	*Airbus* *A340–* *500*	*Boeing* *717–* *200*	*Boeing* *737–* *800*	*Cadair* *Reg. Jet* *100ER*	*Embraer* *EMB-145*	*Fokker* *F70*	*Fokker* *F100*	*Ilyushin* *Il-96M*	*McDon.* */Doug.* *MD-90-30*	*McDon.* */Doug.* *MD-11*	*Tupolev* *Tu-204* *-200*
Take-off (m):															
ISA sea level	2180	2000	2470	3000	3100			1605	1500	1296	1856	3350	2135	2926	2500
ISA +20°C SL	2590	2286	2590	3380	3550		2316			1434	2307			3078	
ISA 5000 ft	2950	3269	3900	4298	4250					1639	2613			3633	
ISA +20°C 5000 ft	4390									1965	3033			4031	
Landing (m):															
ISA sea level	1440	1580	1750	1964	2090	1445	1600	1440	1290	1210	1321	2250	1564	1966	2130
ISA +20°C SL	1440	1580	1750	1964	2090		1600			1210	1321			1966	
ISA 5000 ft	1645	1795	1970	2227	2390					1335	1467			2234	
ISA +20°C 5000 ft	1645	1795	1970	2227	2390					1335	1458			2234	
Speeds (kt/Mach):															
*V*2	143	143	158	158		150				126	136			177	151
*V*app	134	138	135	136	139	130		138	126	119	128			148	
*V*no/*M*mo	350/M0.82	350/M0.82	330/M0.86	330/M0.86	330/M0.86			335/M0.85	320/M0.76	320/M0.77	320/M0.77		0.86	/M0.76	
365/M0.87	314/														
*V*ne/*M*me	381/M0.89	TBD/M0.89		365/M0.93	365/M0.93	365/M0.93					380/M0.84		380/M0.84		
400/M0.92	340/														
*CL*max. (*T/O*)	2.56	3.1	2.21	2.61		2.15				2.16	2.17			2.33	2.32
*CL*max. (*L/D* @ MLM)	3	3.23	2.74	2.89	2.86	3.01		2.1	2.35	2.63	2.59			2.86	

Max cruise:															
Speed (kt)	487	487		500				459	410	461	456	469		M0.87	458
Altitude (ft)	28 000	28 000		33 000			41 000	37 000	37 000	26 000	26 000	9000		31 000	40 000
Fuel consumption (kg/h)	3200	3550		7300					1022	2391	2565			8970	3270
Long range cruise:															
Speed (kt)	448	450	470	475		438	452	424	367	401	414	459	437	M0.81	
Altitude (ft)	37 000	37 000	39 000	39 000		35 000	39 000	37 000	32 000	35 000	35 000	12 000	35 000	31 000	
Fuel consumption (kg/h)	2100	2100		5700			2186.84		880	1475	1716			7060	
Range (nm):															
Max. payload	637	1955	4210	6371	7050				850	1085	1290			5994	1565
Design range	2700	2700	6370	7150	8500	1375	2897	1620	1390	1080	1290	6195	2275	6787	
Max fuel (+ payload)	3672	2602		8089	9000		2927						2267	8234	2079
Design parameters:															
*W/SCL*max	1962.27	2211.48	2269.21	2529.97	2865.71	1811.43		1982	1563	1467	1746			3701	
W/a*CL*to*ST*	2423.85	2590.29	3146.34	4242.69	4144.91	1788.04		2090	1791	1635	2282				
Fuel/pax/nm (kg)	0.0443	0.0465	0.046	0.0472	0.0554	0.0684	0.0465			0.0981	0.0604	0.052	0.0483	0.0543	
Seats × range (seats.nm)	405 000	502 200	1 866 410	2 395 250	2 975 000	145 750	463 520			75 600	138 030	2 075 325	348 075	2 192 201	

Table 10.2 Military aircraft data

Model	*Harrier GR5*	*F–15 Eagle*	*F–14 B*	*MB–339A*	*Hawk T Mk 1*	*Mirage 2000–B*	*F–14D Tomcat*	*Euro-fighter 2000*	*F–117A Stealth*
Date entered service	1969	1972	1974		1976		1990	2001	1982
Role	VTOL attack fighter	Tactical fighter	Shipboard strike fighter (swing wing)	Jet trainer	Jet trainer	Strike fighter	Strike fighter	Air combat fighter	Strike fighter
Contractor	Hawker Siddeley	McDonnel Douglas Corp.	McDonnel Douglas Corp.	Aermacchi	British Aerospace	Dassault Breguet	Grumman	European consortium	Lockheed
Power plant	1 × RR Pegasus turbofan	2 × P&W F100 turbofans with reheat	2 × P&W F400 turbofans with reheat	1 × Piaggio/RR Viper 632–43 turbojet	1 × RR Adour Mk 151	1 × SNECMA M53–5 turbofan with reheat	2 × GE F110–400 turbofans with reheat	2 × Eurojet EJ200 turbofans	2 × GE F404
Thrust (per engine)	9843 kg (21 700 lb)	11 250 kg (25 000 lb)	12 745 kg (28 040 lb)	1814 kg (4000 lb)	2359 kg (5 200 lb)	8790 kg (19 380 lb) with reheat	6363 kg(14 000 lb)	6132 kg(13 490 lb)	
Speed (sea level)	Ma 0.93	Ma 2.5+	Ma 1.2	899 km/h (558 mph)	1037 km/h (645 mph)	Ma 2.3	1997 km/h (1241 mph)	2125 km/h (1321 mph)	High subsonic
Length (m)	14.12	19.43	18.9	10.97	11.85	15.52	19.1	14.5	20.3
Wingspan (m)	9.25	13.06	19.54/11.45	10.25	9.39	8.99	19.55	10.5	13.3
Ceiling (ft)	59 000	65 000	48 000	48 500	48 000	50 000		60 000	
Weight empty	5861 kg (12 922 lb)		18 112 kg (39 850 lb)	3125 kg (6 889 lb)	3628 kg (8 000 lb)	6400 kg (14 080 lb)	18 951 kg (41 780 lb)	9750 kg (21 495 lb)	
Max. take-off weight	13 494 kg (21 700 lb)		33 724 kg (74 192 lb)	5895 kg (13 000 lb)	8330 kg (18 390 lb)	15 000 kg (33 070 lb)	33 724 kg (74 439 lb)	21 000 kg (46 297 lb)	23 625 kg (52 500 lb)

Table 10.2 *Continued*

Model	*A–10 Thunderbolt*	*C 130 Hercules*	*C–5A/B Galaxy*	*B–2 Spirit (Stealth)*	*B–52 Stratofortress*	*B–1B Lancer*	*U–2*	*E–4B*	*TU–95 Bear*
Date entered service	1976	1955	1970	1993	1959	1985	1955	1980	1960
Role	Ground force support	Heavy transport	Strategic airlift	Multi-role heavy bomber	Heavy bomber	Heavy bomber (swing wing)	High altitude reconnaissance aircraft	National Emergency Airborne Command Post	Long-range bomber
Contractor	Fairchild Co.	Lockheed	Lockheed	Northrop	Boeing	Rockwell	Lockheed	Boeing	Tupolev
Power plant	2 × GE TF–34 turbofans	4 × Allison T56 turboprops	4 × GE TF–39 turbofans	4 × GE F–118 turbofans	8 × PW J57 turbojets	4 × GE F–101 turbofans with reheat	1 × PW J75 turbofan	4 × GE CF6 turbofans	4 × Kuznetsov NK–12MV turboprops
Thrust (per engine)	4079 kg (9 065 lb)	3208 kW) 4300 hp	18 450 kg (41 000 lb)	7847 kg (17 300 lb)	6187 kg (13 750 lb)	13 500 kg (29 700 lb) with reheat	7650 kg (17 000 lb)	23 625 kg (52 500 lb)	11 190 kW (15 000 hp)
Speed (sea level)	Ma 0.56	Ma 0.57	Ma 0.72	High subsonic	Ma 0.86	Ma 1.2	Ma 0.57	Ma 0.6	870 km/h (540 mph)
Length (m)	16.16	29.3	75.2	20.9	49	44.8	19.2	70.5	47.48
Wingspan (m)	17.42	39.7	67.9	52.12	56.4	41.8/23.8	30.9	59.7	51.13
Ceiling (ft)	1000	33 000	34 000	50 000	50 000	30 000	70 000	30 000+	20 000+
Weight empty	15909 kg (35 000 lb)				83 250 kg (185 000 lb)	82 250 kg (185 000 lb)			73 483 kg (162 000 lb)
Max. take-off weight	22 950 kg (51 000 lb)		Maximum load capability 130 950 kg (291 000 lb)	152 635 kg (336 500 lb)	219 600 kg (488 000 lb)	214 650 kg (477 000 lb)			170 010 kg (375 000 lb)

10.1.2 Flaps

Trailing and leading edge flaps change the effective camber of the wing, thereby increasing lift. Popular trailing edge types are simple, slotted, double slotted and Fowler flaps (Figure 10.3). Leading edge flaps specifically increase lift at increased angle of incidence and tend to be used in conjunction with trailing edge flaps. Popular types are the simple hinged type and slotted type.

Advanced design concepts such as the *mission adaptive* wing utilize the properties of modern materials in order to flex to adopt different profiles in flight, so separate flaps and slats are not required. Another advanced concept is the *Coanda effect* arrangement, in which turbofan bypass air and exhaust gas is blown onto the upper wing surface, changing the lift characteristics of the wing.

10.1.3 Cabin design

Aircraft cabin design is constrained by the need to provide passenger areas and an underfloor cargo space within the confines of the standard tube-shaped fuselage. This shape of fuselage remains the preferred solution; concept designs with passenger areas enclosed inside a 'flying wing' type body are not yet technically and commercially feasible. Double-deck cabins have been used on a small number of commercial designs but give less facility for cargo carrying, so such aircraft have to be built as a family, incorporating cargo and 'stretch' variants (e.g. the Boeing 747). 'Super-jumbos' capable of carrying 1000+ passengers are currently at the design study stage.

Figure 10.4 shows typical cabin design variants for current airliner models. The objective of any cabin design is the optimization of the payload (whether passengers or freight) within the envelope of a given cabin diameter. Table 10.1 lists comparisons of passenger and freight capabilities for a selection of other aircraft.

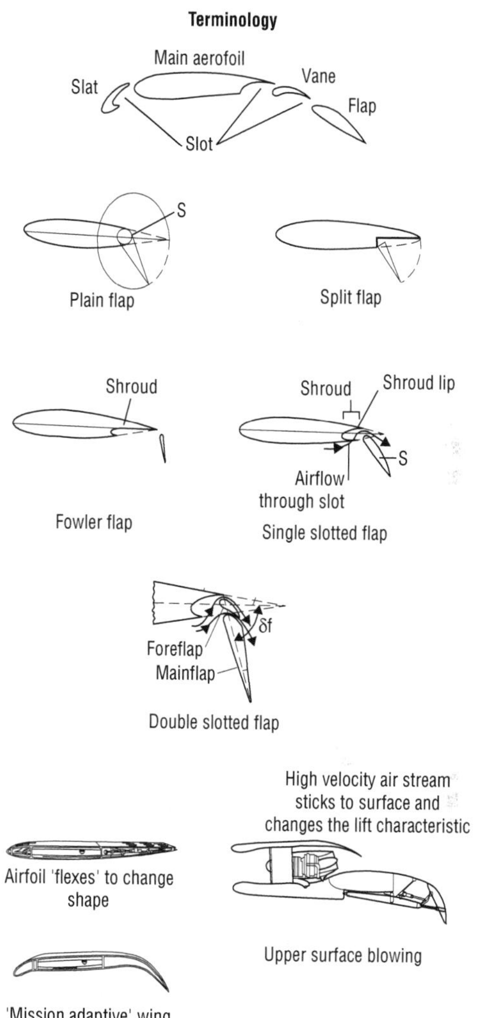

Fig. 10.3 Types of flaps

10.1.4 Ground service capability

Fuselage design is influenced by the ground servicing needs of an aircraft. Ground servicing represents commercial 'downtime' so it is essential to ensure that as many as possible of the ground servicing activities can be carried

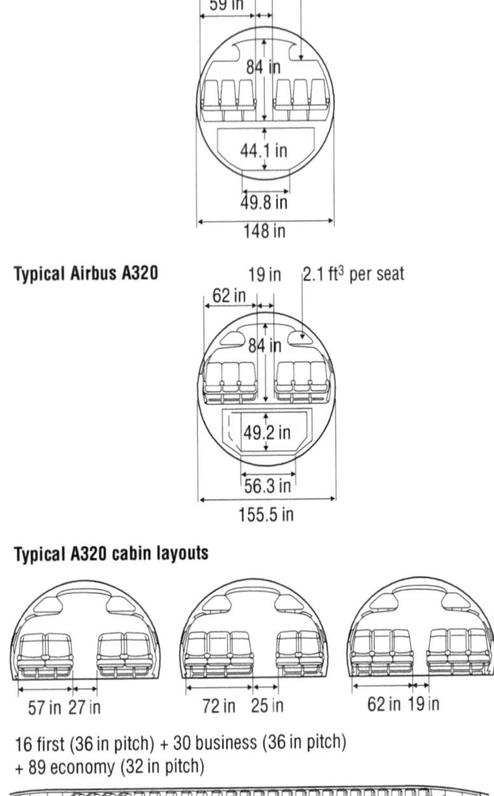

Fig. 10.4 Civil airliner cabin variants

out simultaneously, i.e. the service vehicles and facilities do not get in each others' way. Figure 10.5 shows a general arrangement.

10.1.5 Fuselage construction

Most aircraft have either a monocoque or semi-monocoque fuselage design and use their outer skin as an integral structural or load carrying member. A monocoque (single shell) structure is a thin walled tube or shell which may have stiffening bulkheads or formers installed

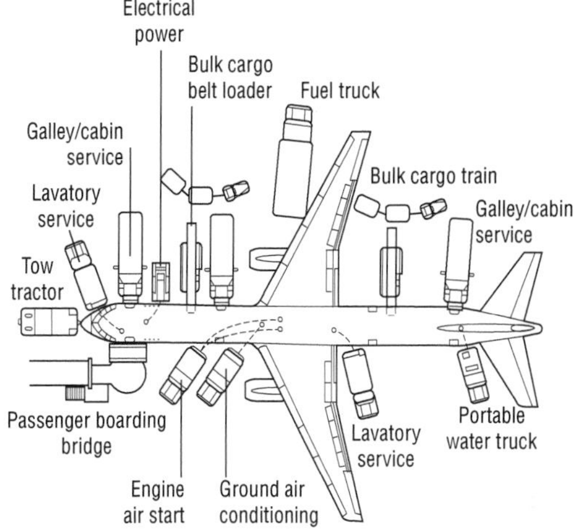

Fig. 10.5 Airliner ground services

within. The stresses in the fuselage are transmitted primarily by the shell. As the shell diameter increases to form the internal cavity necessary for a fuselage, the weight-to-strength ratio changes, and longitudinal stiffeners are added. This progression leads to the semi-monocoque fuselage design which depends primarily on its bulkheads, frames and formers for vertical strength, and longerons and stringers for longitudinal strength. Light general aviation aircraft nearly all have 'stressed-skin' construction. The metal skin exterior is riveted, or bolted and riveted, to the finished fuselage frame, with the skin carrying some of the overall loading. The skin is quite strong in both tension and shear and, if stiffened by other members, can also carry limited compressive load.

10.1.6 Wing construction

General aviation aircraft wings are normally either strut braced or full cantilever type, depending on whether external bracing is used to help transmit loads from the wings to the fuselage. Full cantilever wings must resist all

Table 10.3 Indicative material properties: metallic alloys

	Yield strength		*Ultimate tensile strength*		*Modulus*		*Density*	
	R_m *MN/m²*	F_{tu} *ksi*	R_m *MN/m²*	F_{tu} *ksi*	*E GN/m²*	E_t *psi* × 10^6	ρ *kg/m³*	e_w *lb/in³*
Stainless steel								
15–5 PH forgings	1172.2	170	1310	190	196.5	28.5	7833.44	0.283
17–4 PH sheet	724	105	930.8	135			7861.12	0.284
Alloy steel								
4130 sheet, plate and tube	517.1	75	655	95	200	29	7833.44	0.283
4330 wrought	1282.5	186	1516.9	220	200	29	7833.44	0.283
4340 bar, tube and forging	1482.4	215	1792.7	260	200	29	7833.44	0.283
Heat-resistant steel								
INCONEL 600 sheet, plates, tubes, forgings	206.9	30	551.6	80	206.8	30	8304	0.3
INCONEL 718 sheet plate and tube	999.8	145	1172.1	170	200	29	8304	0.3
Aluminium alloy								
2024-T351 plate	282.7	41	393	57	73.8	10.7	2768	0.1
2024-T4 extrusion	303.4	44	413.7	60	73.8	10.7	2768	0.1
2104-T6 forgings	379.2	55	448.2	65	73.8	10.7	2768	0.1
356-T6 castings	137.9	20	206.9	30	71.7	10.4	2684.96	0.097
Titanium alloy								
6Al–4V sheet, strip plate	999.8	145	1103.2	160	110.3	16	4428.8	0.16
6Al–6V–2Sn forgings	965.3	140	1034.2	150	117.2	17	4539.52	0.164

Table 10.4 Indicative material properties: composites

	Ultimate tensile strength		*Ultimate compressive strength*		*Density*		*Maximum service temperature*	
Material	R_m *MN/m*^{2}	F_{tu} *ksi*	R_c *MN/m*^{2}	F_{cu} *ksi*	ρ *kg/m*^{3}	e_w *lb/in*^{3}	°*C*	°*F*
High temperature epoxy fibreglass	482.6	70	489.5	71	1826.88	0.066	177	350
Phenolic fibreglass	303.4	44	310.3	45	1826.88	0.066	177	350
Epoxy/graphite cloth-woven graphite	551.6	80	586.1	85	1605.44	0.058	177	350
Epoxy/Kevlar cloth	496.5	72	193.1	28	1439.36	0.052	177	350
BMI/graphite	648.1	94	730.9	106	1522.4	0.055	232	450
Polymide graphite	730.9	106	717.1	104	1605.44	0.058	315	600

Table 10.5 General stainless steels – basic data.
Stainless steels are commonly referred to by their AISI equivalent classification (where applicable)

AISI	*Other classifications*	*Type*[2]	*Yield F_{ty} (ksi)*	*[(R_e) MPa]*	*Ultimate F_{tu} (ksi)*	*[(R_m) MPa]*	*E(%) 50 mm*	*HRB*	*%C*	*%Cr*	*% others*[1]
302	ASTM A296 (cast), Wk 1.4300, 18/8, SIS 2331	Austenitic	40	[275.8]	90	[620.6]	55	85	0.15	17–19	8–10 Ni
304	ASTM A296, , Wk 1.4301, 18/8/LC SIS 2333, 304S18	Austenitic	42	[289.6]	84	[579.2]	55	80	0.08	18–20	8–12 Ni
304L	ASTM A351, , Wk 1.4306 18/8/ELC SIS 2352, 304S14	Austenitic	39	[268.9]	80	[551.6]	55	79	0.03	18–20	8–12 Ni
316	ASTM A296, Wk 1.4436 18/8/Mo, SIS 2243, 316S18	Austenitic	42	[289.6]	84	[579.2]	50	79	0.08	16–18	10–14 Ni
316L	ASTM A351, Wk 1.4435, 18/8/Mo/ELC, 316S14, SIS 2353	Austenitic	42	[289.6]	81	[558.5]	50	79	0.03	16–18	10–14 Ni

321	ASTM A240, Wk 1.4541, 18/8/Ti, SIS 2337, 321S18	Austenitic	35	[241.3]	90	[620.6]	45	80	0.08	17–19	9–12 Ni
405	ASTM A240/A276/ A351, UNS 40500	Ferritic	40	[275.8]	70	[482.7]	30	81	0.08	11.5-14.5	1 Mn
430	ASTM A176/A240/ A276, UNS 43000, Wk 1.4016	Ferritic	50	[344.7]	75	[517.1]	30	83	0.12	14–18	1 Mn
403	UNS S40300, ASTM A176/A276	Martensitic	40	[275.8]	75	[517.1]	35	82	0.15	11.5–13	0.5 Si
410	UNS S40300, ASTM A176/A240, Wk 1.4006	Martensitic	40	[275.8]	75	[517.1]	35	82	0.15	11.5-13.5	4.5–6.5 Ni
–	255 (Ferralium)	Duplex	94	[648.1]	115	[793]	25	280 HV	0.04	24–27	4.5–6.5 Ni
–	Avesta SAF 2507[3], UNS S32750	'Super' Duplex 40% ferrite	99	[682.6]	116	[799.8]	≃ 25	300 HV	0.02	25	7 Ni, 4 Mo, 0.3 N

[1]Main constituents only shown.
[2]All austenitic grades are non-magnetic, ferritic and martensitic grades are magnetic.
[3]Avesta trade mark.

loads with their own internal structure. Small, low speed aircraft have straight, almost rectangular, wings. For these wings, the main load is in the bending of the wing as it transmits load to the fuselage, and this bending load is carried primarily by the spars, which act as the main structural members of the wing assembly. Ribs are used to give aerodynamic shape to the wing profile.

10.2 Materials of construction

The main structural materials of construction used in aircraft manufacture are based on steel, aluminium, titanium and composites. Modern composites such as carbon fibre are in increasing use as their mechanical and temperature properties improve. Tables 10.3 and 10.4 show indicative information on the properties of some materials used. Advanced composites can match the properties of alloys of aluminium and titanium but are approximately half their weight. Composite material specifications and performance data are manufacturer specific, and are highly variable depending on the method of formation and lamination. Composite components in themselves are costly to manufacture but overall savings are generally feasible because they can be made in complex shapes and sections (i.e. there are fewer components needing welding, rivets etc.). Some aircraft now have entire parts of their primary structure made of carbon fibre composite. Stainless steel is used for some smaller and engine components. Table 10.5 gives basic data on constituents and properties.

10.2.1 Corrosion

It is important to minimize corrosion in aeronautical structures and engines. Galvanic corrosion occurs when dissimilar metals are in contact in a conducting medium. Table 10.6 shows the relative potentials of pure metals.

Table 10.6 The electrochemical series

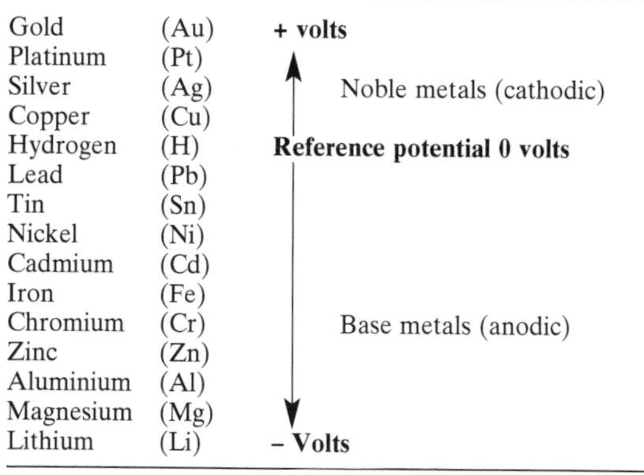

Gold	(Au)	**+ volts**
Platinum	(Pt)	↑
Silver	(Ag)	Noble metals (cathodic)
Copper	(Cu)	
Hydrogen	(H)	**Reference potential 0 volts**
Lead	(Pb)	
Tin	(Sn)	
Nickel	(Ni)	
Cadmium	(Cd)	
Iron	(Fe)	
Chromium	(Cr)	Base metals (anodic)
Zinc	(Zn)	
Aluminium	(Al)	
Magnesium	(Mg)	↓
Lithium	(Li)	**– Volts**

Metals higher in the table become cathodic and are protected by the (anodic) metals below them in the table.

10.3 Helicopter design

10.3.1 Lift and propulsion

Helicopters differ from fixed wing aircraft in that both lift and propulsion are provided by a single item: the rotor. Each main rotor blade acts as slender wing with the airflow producing a high reduction in pressure above the front of the blades, thereby producing lift. Although of high aspect ratio, the blades are proportionately thicker than those of fixed wing aircraft, and are often of symmetric profile. Figure 10.6 shows the principle of helicopter airfoil operation.

10.3.2 Configuration

Figure 10.7 shows the four main configurations used. The most common is the single main and tail rotor type in which the torque of the main rotor drive is counteracted by the lateral force produced by a horizontal-axis tail rotor. Twin tandem rotor machines use intermeshing, counter-rotating rotors with their axes tilted off the vertical to eliminate any torque imparted to the helicopter fuselage. In all designs, lift force is transmitted through the blade roots via the

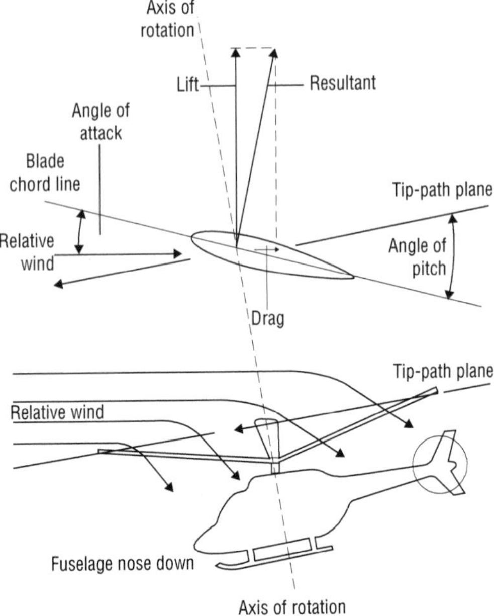

Fig. 10.6 Helicopter principles: lift and propulsion

rotor hub into the main drive shaft, so the helicopter effectively hangs on this shaft.

10.3.3 Forward speed

The performance of standard helicopters is constrained by fixed design features of the rotating rotor blades. In forward flight, the 'retreating' blade suffers reversed flow, causing it to lose lift and stall when the forward speed of the helicopter reaches a certain value. In addition the tip speed of the advancing blades suffers shock-stalls as the blades approach sonic velocity, again causing lift problems. This effectively limits the practical forward speed of helicopters to a maximum of about 310 km/h (192 mph).

10.3.4 Fuel consumption

Helicopters require a higher installed power per unit of weight than fixed wing aircraft. A large proportion of the power is needed simply

Single main and tail rotor
(general purpose helicopter)

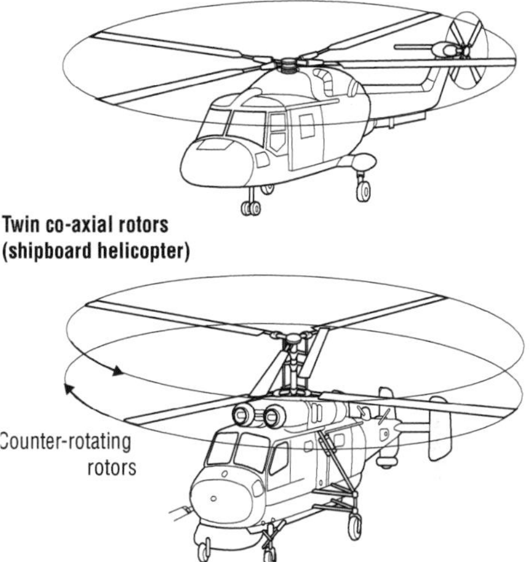

Twin intermeshing rotors

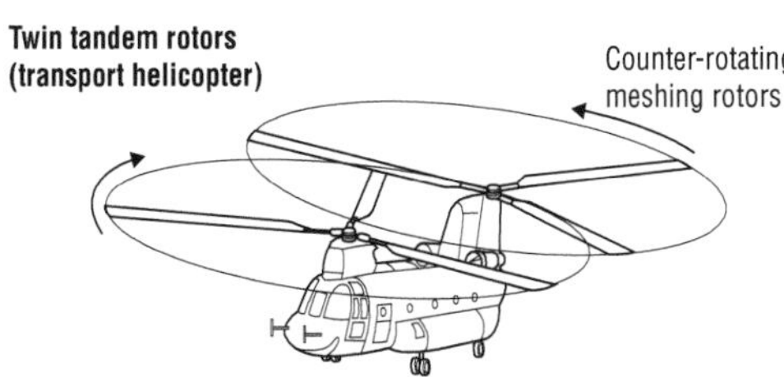

Fig. 10.7 Helicopter configurations

to overcome the force of gravity, and overall specific fuel consumption (sfc) is high. Figure 10.8 shows how sfc is gradually being reduced in commercial helicopter designs.

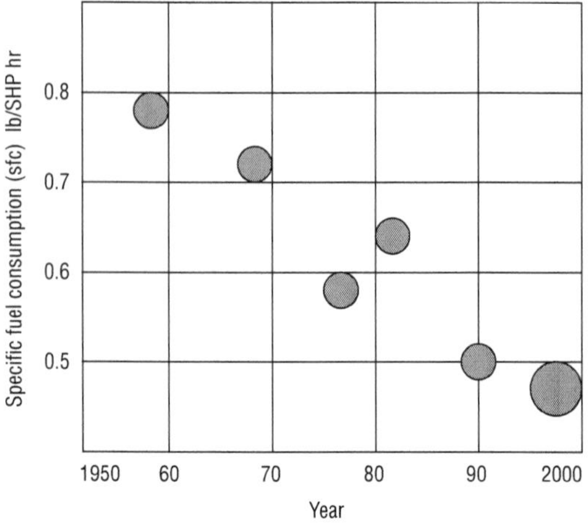

Fig. 10.8 Helicopter sfc trends

10.3.5 Propulsion

Most helicopters are powered either by a single piston engine or by one, two or three gas turbine turboshaft engines. A typical gas turbine model of 1343 kW (1800 hp) comprises centrifugal and axial compressor stages and two stage 'free power' turbine. The largest units in use are the 8500 kW+ (11400 hp+) 'Lotarev' turboshafts used to power the Mil-26 heavy transport helicopter. Table 10.7 shows comparative data from various manufacturers' designs.

10.4 Helicopter design studies

Helicopter design studies follow the general pattern shown in Figure 10.9. The basis of the procedure is to start with estimates of gross weight and installed power based on existing helicopter designs. First estimates also have to be made for disc loading and forward flight drag. The procedure is then interative (as with the fixed wing design study outlined in Chapter 9) until a design is achieved that satisfies all the design requirements.

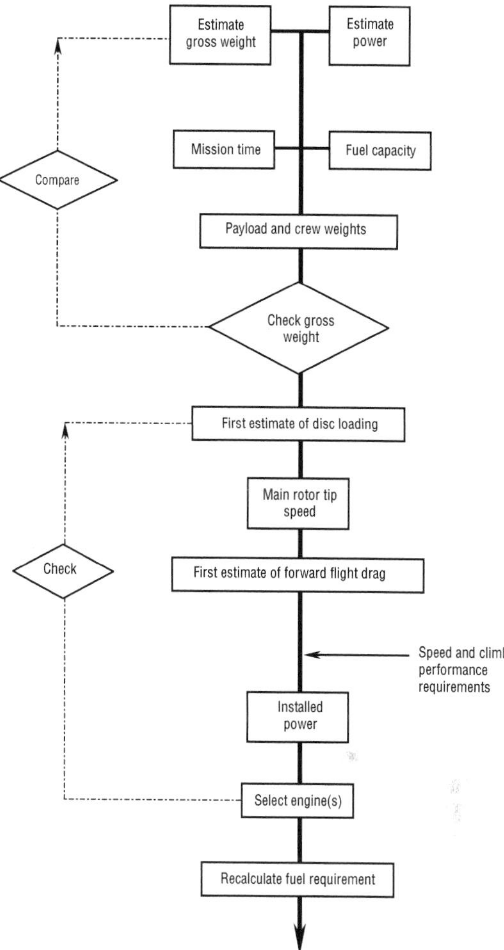

Fig. 10.9 Helicopter design studies: the basic steps

10.4.1 Helicopter operational profile

For military helicopters, the operational profile is frequently termed *mission capability*. The relatively short range and low endurance of a helicopter, compared to fixed wing aircraft, means that the desired mission profile has a significant influence on the design. Figure 10.10 shows a typical military mission profile.

Table 10.7 Helicopter comparisons

Model	*Type*	*Entered service*	*Engines*			*Weight*		*Performance*	
			No.	*Type*	*Power (each)*	*Empty*	*Max. loaded*	*Max. speed at sea level*	*Max. rate of climb*
Aerospatiale SA 330 Puma	Medium transport	1965	2	Turbomeca turboshaft	991 kW (1328 hp)	3536 kg (7795 lb)	6400 kg (14 110 lb)	280 km/h (174 mph)	366 m/min (1200 ft/min)
Agusta A129 Mangusta	Attack helicopter	1983	2	GEM 2 turboshaft	708 kW (952 hp)	2529 kg (5575 lb)	4100 kg (9039 lb)	259 km/h (161 mph)	637 m/min (2090 ft/min)
Bell Huey AH–1 Cobra	Attack helicopter	1965	1	turboshaft	1044 kW (1400 hp)	2755 kg (6073 lb)	4310 kg (9500 lb)	277 km/h (172 mph	375 m/min (1230 ft/min)
Eurocopter UHU/HAC	Anti-tank helicopter	1991	2	MTR turboshaft	1160 kW (1556 hp)	3300 kg (7275 lb)	5800 kg (12 787 lb)	280 km/h (174 mph)	600 m/min (1970 ft/min)
Kamov Ka–50	Close-support helicopter	1982	2	Klimov turboshaft	1634 kW (2190 hp)	4550 kg (10 030 lb)	10 800 kg (23 810 lb	310 km/h (193 mph)	600 m/min (1970 ft/min)

Mil Mi–26	Heavy transport helicopter	1979	2	Lotaren turboshaft	8504 kW (11 400 hp)	28 200 kg (62 169 lb)	49 500 kg (10 9127 lb)	295 km/h (183 mph)	–
Boeing CH–47 Chinook	Medium transport helicopter	1961	2	Allied signal turboshaft	1641 kW (2200 hp)	9242 kg (20 378 lb)	20 866 kg (46 000 lb)	306 km/h (190 mph)	878 m/min (2880 ft/min)
Bell/Boeing V–22 Osprey	Multi-role VTOL rotorcraft	1989	2	Allison turboshaft	4588 kW (6150 hp)	14 800 kg (32 628 lb)	VTOL: 21546 kg (47 500 lb) STOL: 24 948 kg (5500 lb)	629 km/h (391 mph)	–
EH101 Merlin	Multi-role helicopter	1987	3	GE turboshaft	1522 kW (2040 hp)	9072 kg (20 000 lb)	14 600 kg (32 188 lb)	309 km/h (192 mph)	–

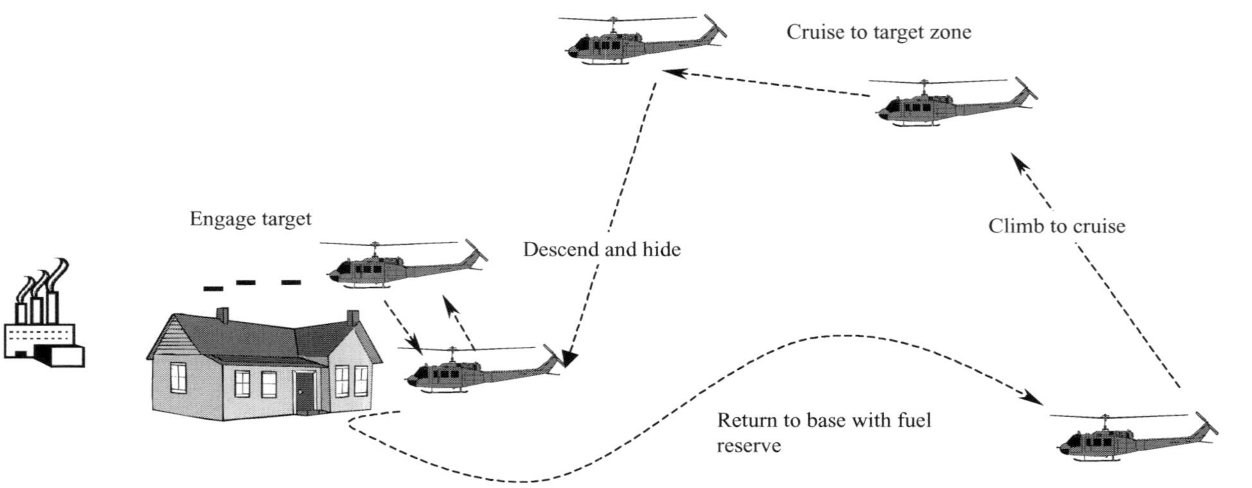

Fig. 10.10 Typical military helicopter 'mission profile'

Section 11

Airport design and compatibility

Airports play an important role in the civil and military aeronautical industries. They are part of the key infrastructure of these industries and, because of their long construction times and high costs, act as one of the major fixed *constraints* on the design of aircraft.

11.1 Basics of airport design

11.1.1 The airport design process

The process of airport design is a complex compromise between multiple physical, commercial and environmental considerations. Physical facilities needed include runways, taxiways, aprons and strips, which are used for the landing and take-off of aircraft, for the manoeuvring and positioning of aircraft on the ground, and for the parking of aircraft for loading and discharge of passengers and cargo. Lighting and radio navigation are essential for the safe landing and take-off of aircraft. These are supplemented by airfield markings, signals, and air traffic control facilities. Support facilities on the airside include meteorology, fire and rescue, power and other utilities, maintenance, and airport maintenance. Landside facilities are the passenger and cargo terminals and the infrastructure system, which includes parking, roads, public transport facilities, and loading and unloading areas. At all stages of the design process, the issue of *aircraft compatibility* is of prime importance – an airport must be suitable for the aircraft that will use it, and vice versa.

11.1.2 Airport site selection

The airport site selection process includes several stages of activity. Table 11.1 shows the main 'first stage balance factors'.

Table 11.1 Airport site selection: 'first stage balance factors'

Aeronautical requirements	*Environmental constraints*
• Flat area of land (up to 3000* acres for a large facility) • Sufficiently close to population centres to allow passenger access	• Should not impinge on areas of natural beauty • Sufficiently far away from urban centres to minimize the adverse effects of noise etc.

*Note: Some large international airports exceed this figure (e.g. Jeddah, Saudi Arabia and Charles de Gaulle, Paris).

11.1.3 Operational requirements – 'rules of thumb'

There is a large variation in the appearance and layout of airport sites but all follow basic 'rules of thumb':

- The location and orientation of the runways are primarily decided by the requirement to avoid obstacles during take-off and landing procedures. 15 km is used as a nominal 'design' distance.
- Runway configuration is chosen so that they will have manageable crosswind components (for the types of aircraft being used) for at least 95% of operational time.
- The number of runways available for use at any moment determines the *operational capacity* of the airport. Figure 11.1 shows common runway layouts. Crosswind facility is achieved by using either a 'crossed' or 'open or closed vee' layout.
- Operational capacity can be reduced under IFR (Instrument Flying Rules) weather conditions when it may not be permissible to use some combinations of runways simultaneously unless there is sufficient separation (nominally 1500+ metres).

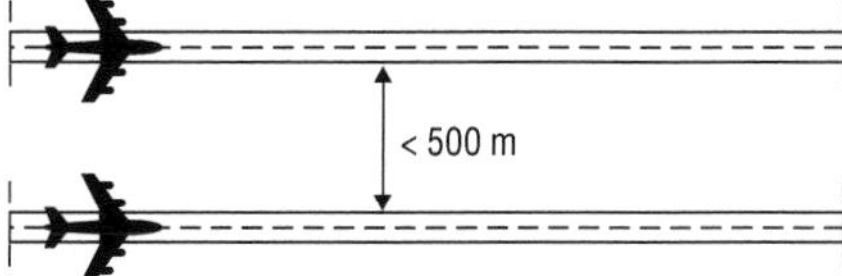

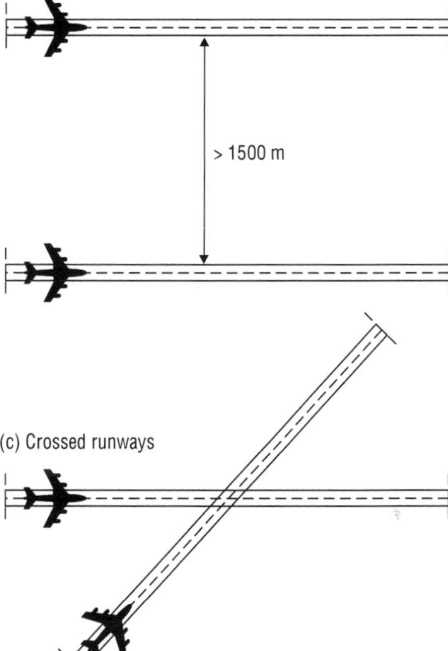

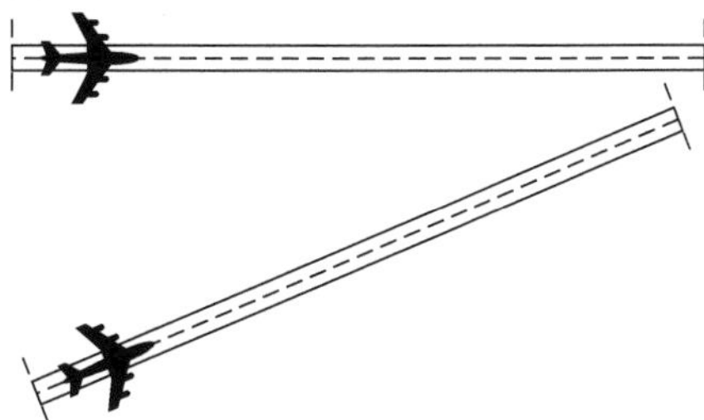

Fig. 11.1 Common runway layouts

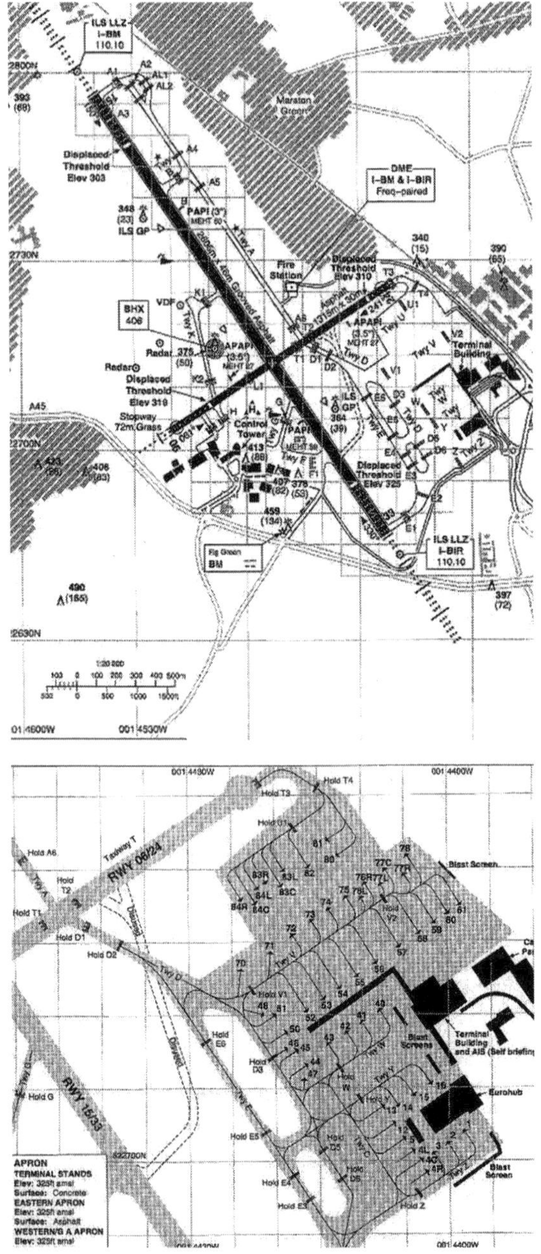

Fig. 11.2 Birmingham airport – a crossed runway layout

Figure 11.2 shows Birmingham (UK) airport layout – a mid-size regional airport with crossed runway design. Figure 11.3 shows a large national airport with a crossed and independent parallel runway layout.

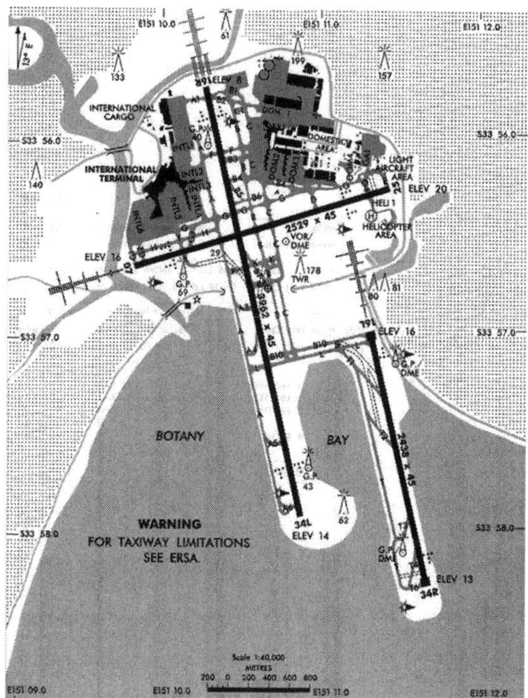

Fig. 11.3 A crossed and independent parallel runway layout

11.1.4 Aircraft:airport compatibility

A prime issue in the design of a new airport, or the upgrading of an existing one, is aircraft:airport compatibility. Aircraft and airport design both have long lead times, which means that new airports have to be designed to meet the constraints of existing and planned aircraft designs, and vice versa. These constraints extend across the various elements of airport design, i.e. runway length, width and

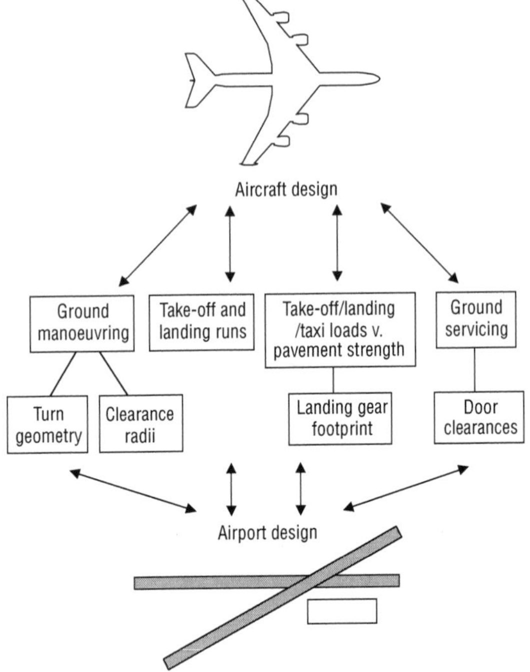

Fig. 11.4 Aircraft:airport compatibility – some important considerations

orientation, taxiways and holding bays, pavement design, ground servicing arrangements and passenger/cargo transfer facilities. Figure 11.4 shows a diagrammatic representation of the situation.

Details of aircraft characteristics are obtained from their manufacturers' manuals, which address specifically those characteristics which impinge upon airport planning. The following sections show the typical format of such characteristics, using as an example the Boeing 777 aircraft.

General dimensions

The general dimensions of an aircraft have an influence on the width of runways, taxiways, holding bays and parking bays. Both wingspan

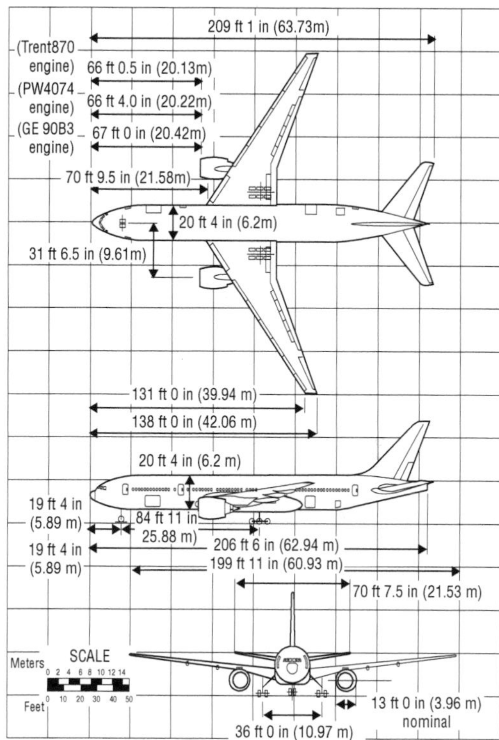

Fig. 11.5 Aircraft:airport compatibility – general dimensions. Figure shows Boeing 777-200. Courtesy Boeing Commercial Airplane Group

and overall length can place major constraints on an airport's design. Figure 11.5 shows typical data.

General clearances

Aircraft ground clearance is an important criterion when considering ground-based obstacles and both fixed and mobile ground servicing facilities. Figure 11.6 shows typical data.

Door location and type

The location and type of doors have an influence on passenger access and cargo handling design aspects of the overall airport facility.

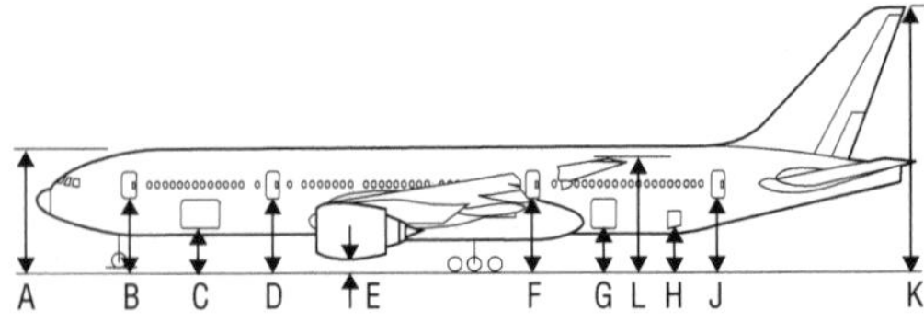

	Minimum*		Maximum*	
	Feet - inches	Meters	Feet - inches	Meters
A	27-6	8.39	28-6	8.68
B	15-5	4.71	16-5	5.00
C	9-3	2.81	10-0	3.05
D	16-0	4.88	16-7	5.07
E (PW)	3-2	0.96	3-5	1.04
E (GE)	2-10	0.85	3-1	0.93
E (RR)	3-7	1.09	3-10	1.17
F	16-10	5.14	17-4	5.28
G (Large door)	10-7	3.23	11-2	3.41
G (Small door)	10-6	3.22	11-2	3.40
H	10-7	3.23	11-5	3.48
J	17-4	5.28	18-2	5.54
K	60-5	18.42	61-6	18.76
L	23-6	7.16	24-6	7.49

Fig. 11.6 Aircraft:airport compatibility – ground clearances. Figure shows Boeing 777-200. Courtesy Boeing Commercial Airplane Group

Figures 11.7 and 11.8 show typical passenger door locations and clearances. Figures 11.9 and 11.10 show comparable data for cargo doors.

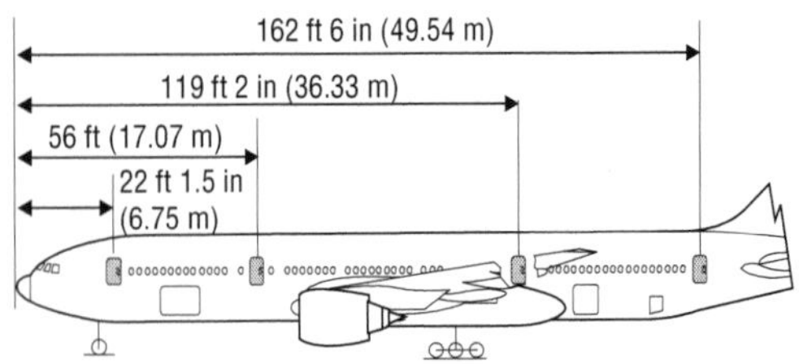

Fig. 11.7 Aircraft:airport compatibility – passenger door locations. Figure shows Boeing 777-200. Courtesy Boeing Commercial Airplane Group

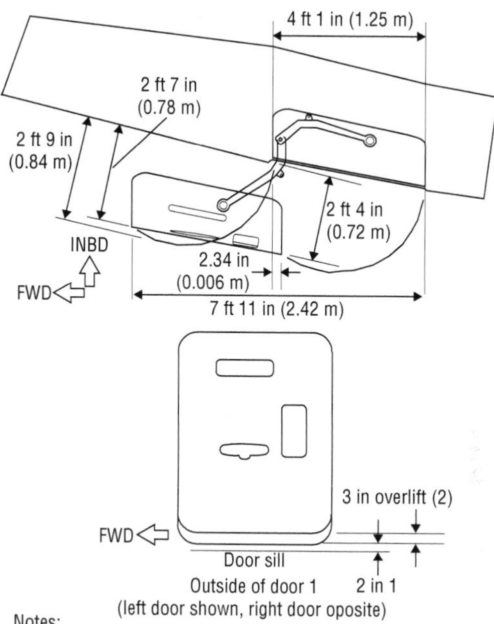

Fig. 11.8 Aircraft:airport compatibility – passenger door clearances. Figure shows Boeing 777-200. Courtesy Boeing Commercial Airplane Group

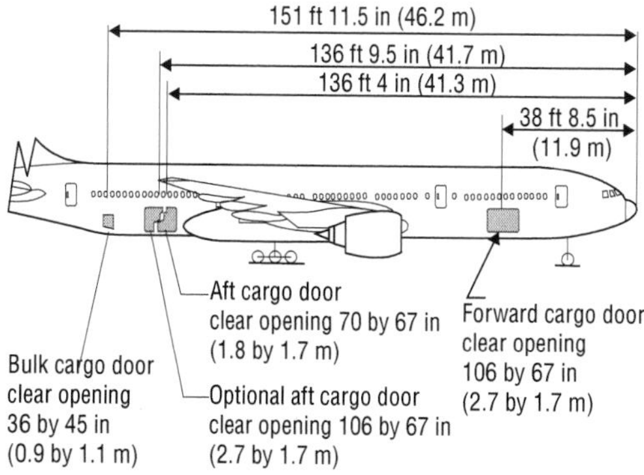

Fig. 11.9 Aircraft:airport compatibility – cargo door locations. Figure shows Boeing 777-200. Courtesy Boeing Commercial Airplane Group

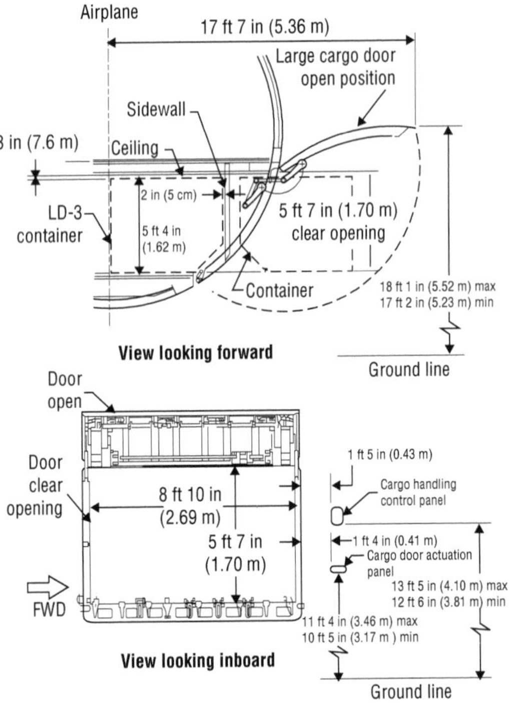

Fig. 11.10 Aircraft:airport compatibility – cargo door clearances. Figure shows Boeing 777-200. Courtesy Boeing Commercial Airplane Group

Runway take-off and landing length requirements

Every aircraft manual contains runway length requirements for take-off and landing. A series of characteristic curves are provided for various pressure altitudes (i.e. the airport location above sea level), ambient temperature aircraft weights, wind, runway gradient and conditions etc. Figures 11.11 and 11.2 show typical data, and the way in which the graphs are presented.

Manoeuvring geometry and clearances

Aircraft turn radii and clearances can influence the design of taxiways, holding bays intersections etc. as well as parking bays and manoeuvring

Notes:
- Consult using airline for specific operating procedure prior to facility design
- Zero runway gradient
- Zero wind

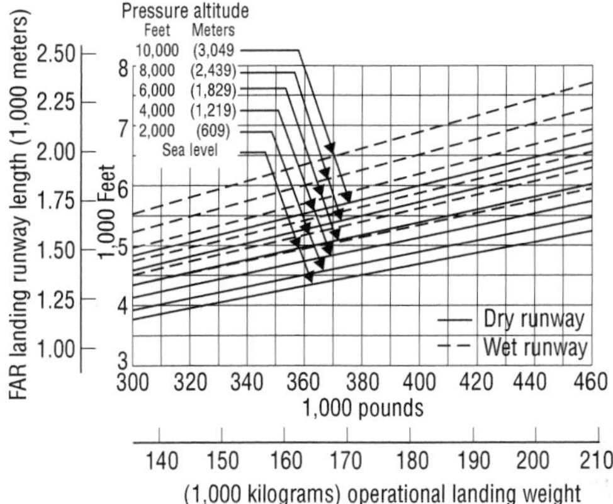

Fig. 11.11 Aircraft:airport compatibility – landing runway length requirements. Figure shows Boeing 777-200. Courtesy Boeing Commercial Airplane Group

Notes:
- Consult using airline for specific operating procedure prior to facility design
- Air conditioning off
- Zero runway gradient
- Zero wind

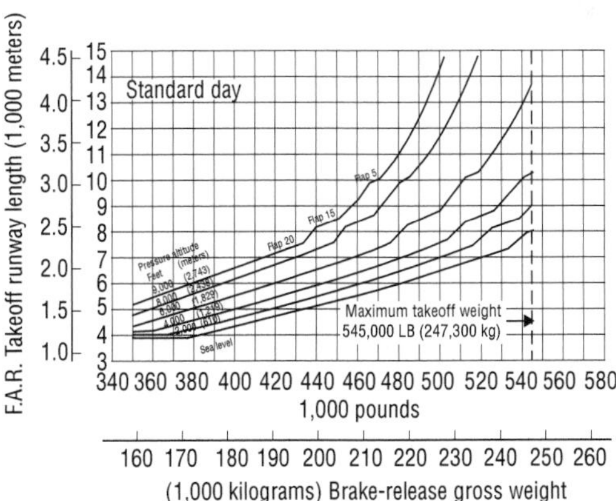

Fig. 11.12 Aircraft:airport compatibility – take-off runway length requirements. Figure shows Boeing 777-200. Courtesy Boeing Commercial Airplane Group

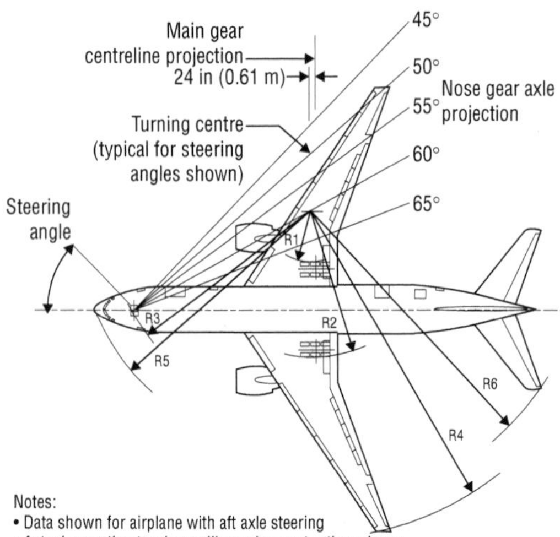

Notes:

- Data shown for airplane with aft axle steering
- Actual operating turning radii may be greater than shown.
- Consult with airline for specific operating procedure
- Dimensions rounded to nearest foot and 0.1 meter

Steering angle	R1 Inner gear		R2 Outer gear		R3 Nose gear	
(Deg)	Ft	M	Ft	M	Ft	M
30	123	37.5	165	50.3	168	51.3
35	98	29.7	140	42.6	147	44.8
40	78	23.7	120	36.6	131	40.0
45	62	18.9	104	31.7	120	36.4
50	49	14.8	91	27.7	111	33.7
55	37	11.2	79	24.1	103	31.5
60	27	8.1	69	21.0	98	29.9
65	17	5.3	60	18.2	94	28.6
70 (max)	9	2.7	51	15.6	90	27.6

	R4 Wing tip		R5 Nose		R6 Tail	
	Ft	M	Ft	M	Ft	M
	247	75.3	177	53.8	209	63.6
	222	67.6	157	47.8	187	57.1
	202	61.7	142	43.4	171	52.2
	187	56.9	132	40.2	159	48.5
	174	52.9	124	37.7	150	45.6
	162	49.5	118	35.8	142	43.2
	152	46.5	113	34.4	135	41.2
	143	43.7	109	33.3	130	39.5
	135	41.2	107	32.5	125	38.1

Fig. 11.13 Aircraft:airport compatibility – turning radii. Figure shows Boeing 777-200. Courtesy Boeing Commercial Airplane Group

capabilities in the vicinity of passenger and cargo loading facilities. Different types and sizes of aircraft can have very different landing gear tracks and 'footprints' – hence an airport's design often has to incorporate compromises, so that it is suitable for a variety of aircraft types. Figure 11.13 shows the typical way that turn radii are

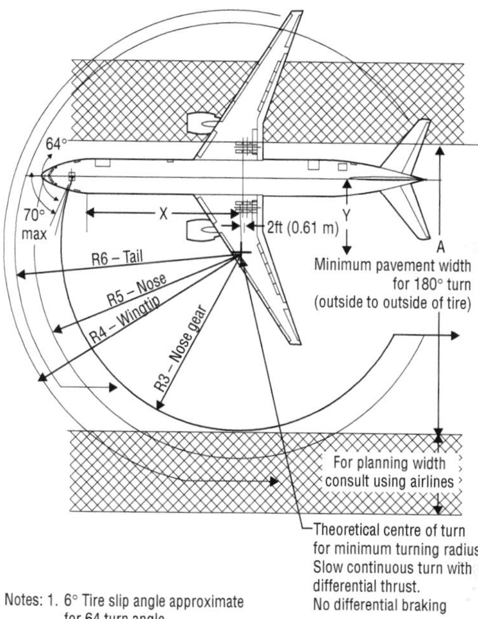

Notes: 1. 6° Tire slip angle approximate for 64 turn angle.
2. Consult using airline for specific operating procedure.
3. Dimensions are rounded to the nearest foot and 0.1 meter.

Airplane model	Effective steering angle (Deg)	X		Y		A		R3	
		FT	M	FT	M	FT	M	FT	M
777-200	64	83	5.3	40	12.2	156	47.5	95	29.0
777-300	64	100	30.6	49	14.9	182	55.4	112	34.0

R4		R5		R6	
FT	M	FT	M	FT	M
145	44.2	110	33.5	131	39.9
154	46.8	129	39.4	149	45.3

Fig. 11.14 Aircraft:airport compatibility – clearance radii. Figure shows Boeing 777-200. Courtesy Boeing Commercial Airplane Group

expressed. Figure 11.14 shows corresponding clearance radii and the way in which the aircraft characteristics for a 180° turn define the minimum acceptable pavement width that is necessary.

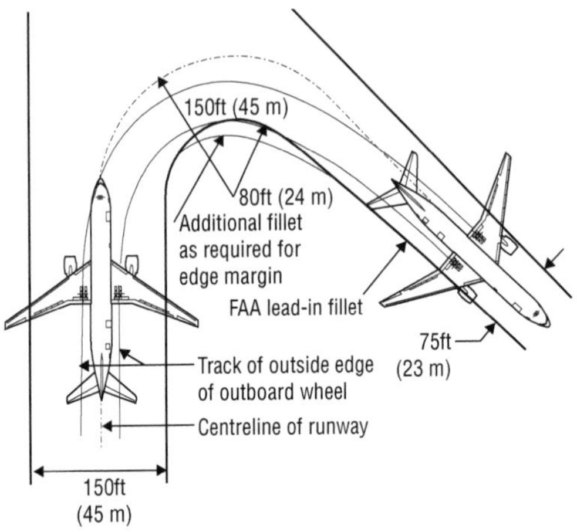

Fig. 11.15 Aircraft:airport compatibility – runway and taxiway intersections (> 90°). Figure shows Boeing 777-200/300. Courtesy Boeing Commercial Airplane Group

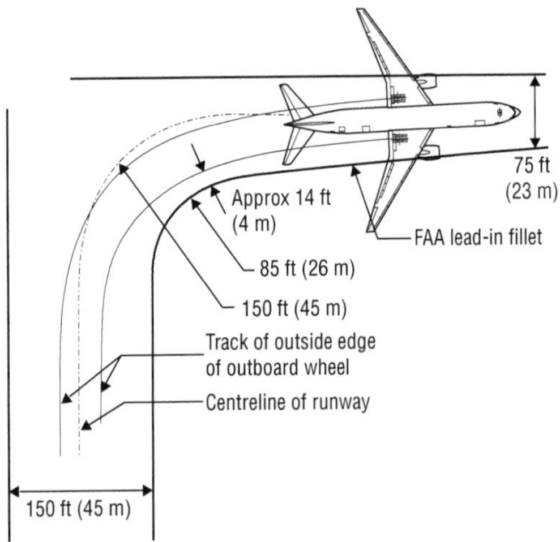

Fig. 11.16 Aircraft:airport compatibility – runway and taxiway intersections (90°). Figure shows Boeing 777-200/300. Courtesy Boeing Commercial Airplane Group

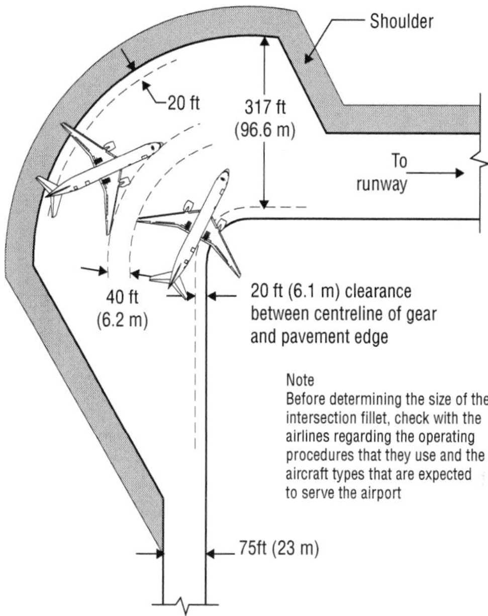

Fig. 11.17 Aircraft:airport compatibility – holding bay sizing. Figure shows Boeing 777-200/300. Courtesy Boeing Commercial Airplane Group

An important aspect of aircraft:airport compatibility is the required geometry of runway and taxiway turnpaths and intersections. Consideration must be given to features such as *intersection fillets*, sized to accommodate aircraft types expected to use the airport. Figures 11.15 and 11.16 show typical characteristics for 90° and > 90° turnpaths. Figure 11.17 shows a corresponding holding bay arrangement – note the need for adequate wing tip clearance between holding aircraft, and clearance between each aircraft's landing gear track and the pavement edge.

Pavement strength

Airports' pavement type and strength must be designed to be compatible with the landing gear loadings, and the frequency of these loadings, of the aircraft that will use it. A standardized

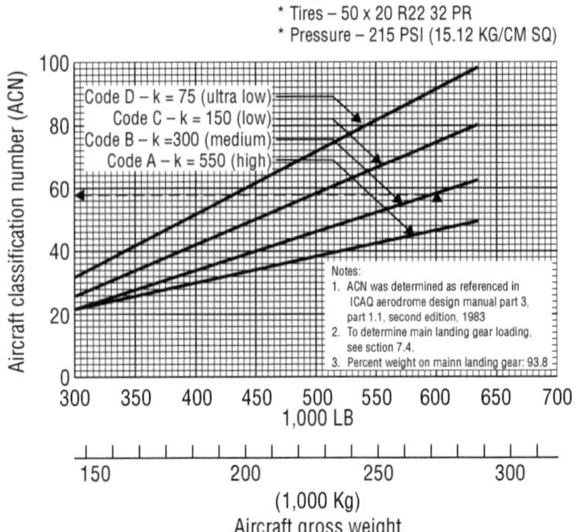

Fig. 11.18 Aircraft:airport compatibility – aircraft classification No.: rigid pavement. Data for Boeing 777-200. Courtesy Boeing Commercial Airplane Group

compatibility assessment is provided by the Aircraft Classification Number/Pavement Classification Number (ACN/PCN) system. An aircraft having an ACN equal to or less than the pavement's PCN can use the pavement safely, as long as it complies with any restrictions on the tyre pressures used. Figures 11.18 and 11.19 show typical rigid pavement data (see also Section 11.2) whilst Figure 11.20 shows data for flexible pavement use.

Airside and landside services

The main airside and landside services considered at the airport design stage are outlined in Table 11.2.

11.1.5 Airport design types

The design of an airport depends principally on the passenger volumes to be served and the type of passenger involved. Some airports have a very high percentage of passengers who are transiting the airport rather than treating it as their final destination, e.g. Chicago O'Hare

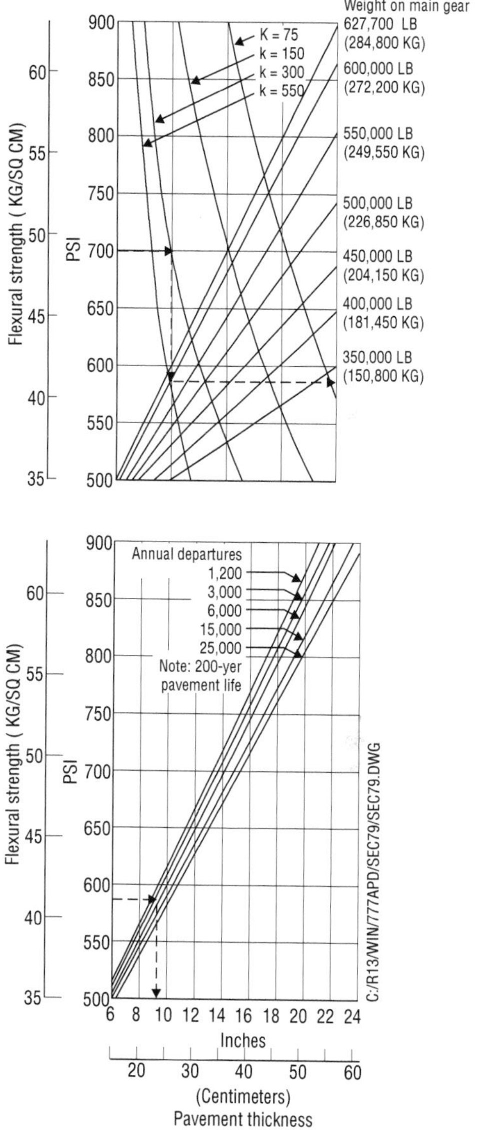

Fig. 11.19 Aircraft:airport compatibility – rigid pavement requirements. Data for Boeing 777-200. Courtesy Boeing Commercial Airplane Group

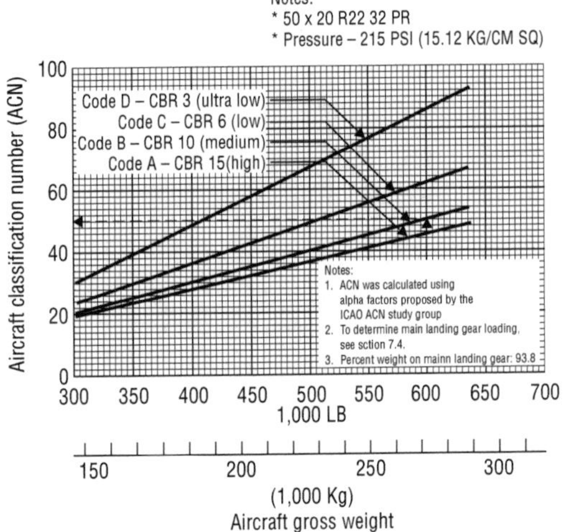

Fig. 11.20 Aircraft:airport compatibility – aircraft classification No.: flexible pavement. Data for Boeing 777-200. Courtesy Boeing Commercial Airplane Group

International (USA). These are referred to as *hubbing* airports. At a hub, aircraft from a carrier arrive in waves, and passengers transfer between aircraft during the periods when these waves are on the ground. By using a hub-and-spoke design philosophy, airlines are able to increase the load factors on aircraft and to provide more frequent departures for passengers – at the cost, however, of inconvenient interchange at the hub.

11.1.6 Airport capacity

The various facilities at an airport are designed to cope adequately with the anticipated flow of passengers and cargo. At smaller single-runway airports, limits to capacity usually occur in the terminal areas, since the operational capacity of a single runway with adequate taxiways is quite large. When passenger volumes reach approximately 25 million per year, a single runway is no longer adequate to handle the number of aircraft movements that take place during peak periods. At this point at least one additional runway,

Table 11.2 Airside and landside service considerations

Landside	*Airside*
• Ground passenger handling including: – Check-in – Security – Customs and immigration – Information – Catering – Cleaning and maintenance – Shopping and concessionary facilities – Ground transportation • Management and administration of airport staff	• Aircraft apron handling • Airside passenger transfer • Baggage and cargo handling • Aircraft fuelling • Cabin cleaning and catering • Engine starting maintenance • Aircraft de-icing • Runway inspection and maintenance • Firefighting and emergency services • Air traffic control

Other basic airport requirements are:

- **Navigation aids** – normally comprising an Instrument Landing System (ILS) to guide aircraft from 15 miles from the runway threshold. Other commonly installed aids are:
  - Visual approach slope indicator system (VASIS)
  - Precise approach path indicator (PAPI)
- **Airfield lighting** – White neon lighting extending up to approximately 900 m before the runway threshold, threshold lights (green), 'usable pavement end' lights (red) and taxiway lights (blue edges and green centreline).

permitting simultaneous operation, is required. Airports with two simultaneous runways can frequently handle over 50 million passengers per year, with the main constraint being, again, the provision of adequate terminal space.

Layouts with four parallel runways can have operational capacities of more than one million aircraft movements per year and annual passenger movements in excess of 100 million. The main capacity constraints of such facilities are in the provision of sufficient airspace for controlled aircraft movements and in the provision of adequate access facilities. Most large international airport designs face access problems before they reach the operational capacity of their runways.

11.1.7 Terminal designs

Open apron and linear designs

The simplest layout for passenger terminals is the *open apron design* (Figure 11.21(a)) in which aircraft park on the apron immediately adjacent to the terminal and passengers walk across the apron to board the aircraft. Frequently, the aircraft manoeuvre in and out of the parking

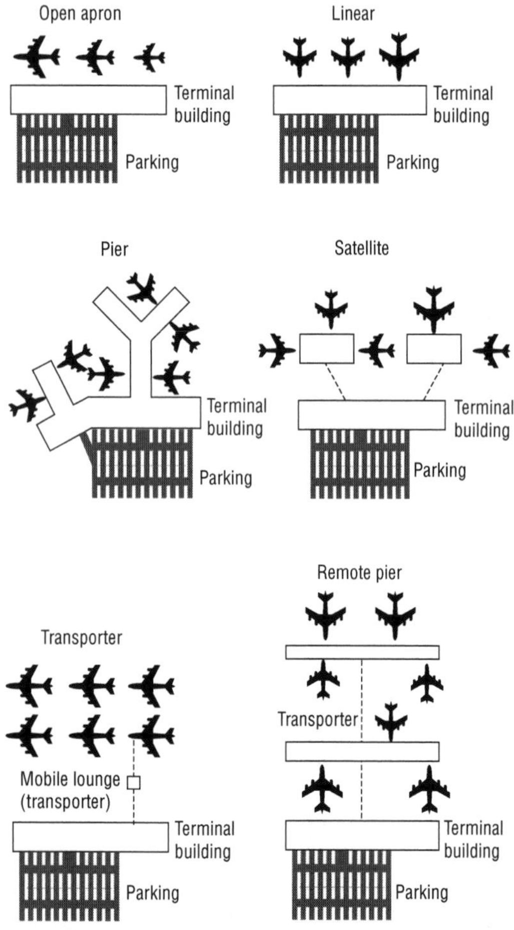

Fig. 11.21 Airport terminal designs

positions under their own power. When the number of passengers walking across the apron reaches unmanageable levels the optimum design changes to the *linear type* (Figure 11.21(b)) in which aircraft are parked at gates immediately adjacent to the terminal itself, and passengers board by air bridge. The limitation of the linear concept is usually the long building dimensions required; this can mean long walking distances for transferring passengers and other complications related to building operation. In most designs, building lengths reach a maximum of approximately 700 m. Examples are Kansas City International, USA, Munich, Germany (Figure 11.22), and Paris Charles de Gaulle, France.

Pier and satellite designs

The *pier concept* (Figure 11.21(c)) has a design philosophy in which a single terminal building serves multiple aircraft gates (Frankfurt and Schipol used this concept prior to their recent expansion programmes). The natural extension of this is the *satellite concept* (Figure 11.21(d)), in which passengers are carried out to the satellites by automated people-mover or automatic train. This design is difficult to adapt to the changing size of aircraft and can be wasteful of apron space.

Transporter designs

The *transporter concept* (Figure 11.21(e)) is one method of reducing the need for assistance for aircraft manoeuvring on the apron and eliminating the need for passengers to climb up and down stairways to enter or exit the aircraft. Passengers are transported directly to the aircraft by specialized transporter vehicles which can be raised and lowered (Dulles International, USA and Jeddah's King Abdul Aziz International Airport, Saudi Arabia, are examples).

Remote pier designs

In this design (Figure 11.21(f)) passengers are brought out to a remote pier by an automatic

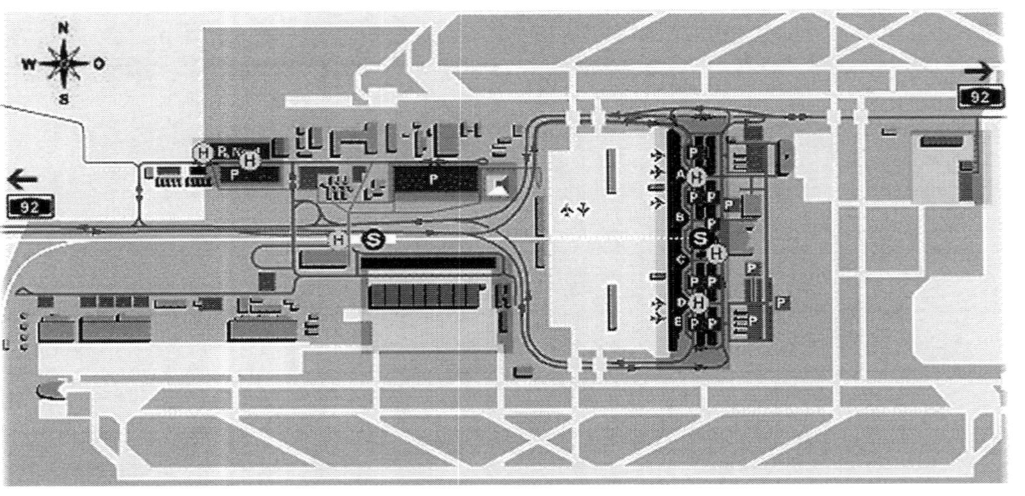

Fig. 11.22 Munich airport layout – a 'linear' design

people-mover and embark or disembark in the conventional manner (Stansted, UK, is an example).

Unit terminals

The term *unit terminal* is used when an airport passenger terminal system comprises more than one terminal. Unit terminals may be made up of a number of terminals of similar design (Dallas-Fort Worth, USA), terminals of different design (London Heathrow), terminals fulfilling different functions (London Heathrow, Arlanda, Stockholm), or terminals serving different airlines (Paris Charles de Gaulle). The successful operation of unit terminal airports requires rapid and efficient automatic people-movers that operate between the terminals.

11.1.8 The apron

An important requirement in the design of an airport is minimizing the time needed to service an aircraft after it has landed. This is especially important in the handling of short-haul aircraft, where unproductive ground time can consume an unacceptably large percentage of flight time. The turnaround time for a large passenger transport between short-haul flights can be as little as 25 minutes. During this period, a large number of service vehicles circulate on the apron (see Figure 10.5 in Chapter 10), so an important aspect of the efficient operation of an airport facility is the marshalling of ground service vehicles and aircraft in the terminal apron area. Such an operation can become extremely complex at some of the world's busiest international airports, where an aircraft enters or leaves the terminal apron approximately every 20 seconds.

11.1.9 Cargo facilities

Although only approximately 1–2% of worldwide freight tonnage is carried by air, a large international airport may handle more than one million tons of cargo per year. Approximately 10% of air cargo is carried loose or in bulk, the remainder in air-freight containers. In developed countries, freight is moved by mobile mechanical equipment such as stackers, tugs, and forklift trucks. At high-volume facilities, a mixture of mobile equipment and complex fixed stacking and movement systems must be used. Fixed systems are known as transfer vehicles (TVs) and elevating transfer vehicles (ETVs). An area of high business growth is specialized movement by courier companies which offer door-to-door delivery of small packages at premium rates. Cargo terminals for the small-package business are designed and constructed separately from conventional air-cargo terminals – they operate in a different manner, with all packages being cleared on an overnight basis.

11.2 Runway pavements

Modern airport runway lengths are fairly static owing to the predictable take-off run requirements of current turbofan civil aircraft. All but the smallest airports require pavements for runways, taxiways, aprons and maintenance areas. Table 11.3 shows basic pavement requirements and Figure 11.23 the two common types.

Table 11.3 Runway pavements – basic requirements

- Ability to bear aircraft weight without failure
- Smooth and stable surface
- Free from dust and loose particles
- Ability to dissipate runway loading without causing subgrade/subsoil failure
- Ability to prevent weakening of the subsoil by rainfall and frost intrusion

The two main types of pavement are:

- **Rigid pavements:** Cement slabs over a granular sub-base or sub-grade. Load is transmitted mainly by the distortion of the cement slabs.
- **Flexible pavements:** Asphalt or bitumous concrete layers overlying granular material over a prepared sub-grade. Runway load is spread throughout the depth of the concrete layers, dissipating sufficiently so the underlying subsoil is not overloaded.

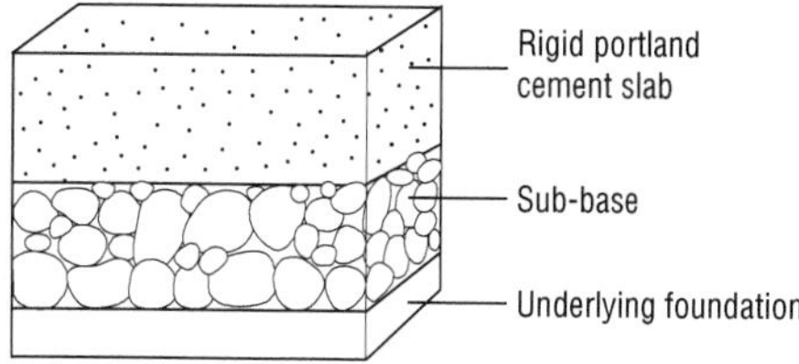

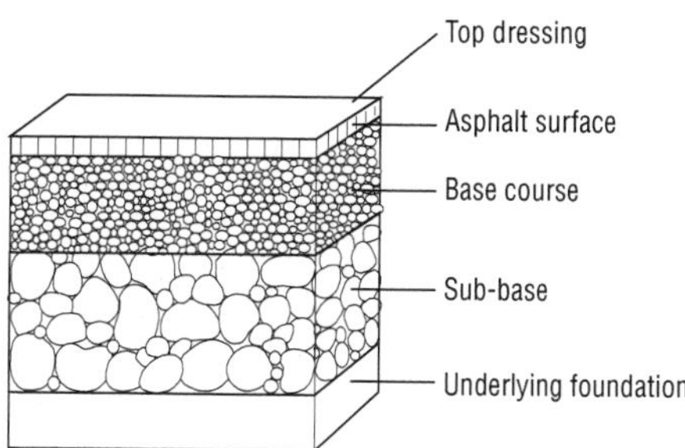

Fig. 11.23 Rigid and flexible runway pavements

11.3 Airport traffic data

Tables 11.4 and 11.5 show recent traffic ranking data for world civil airports.

11.4 FAA–AAS Airport documents

Technical and legislative aspects of airport design are complex and reference must be made to up-to-date documentation covering this subject. The Office of Airport Safety and Standards (ASS) serves as the principal organization of United States Federal Aviation Authority (FAA) responsible for all airport programme matters about standards for airport design, construction, maintenance, operations and safety. References available are broadly as shown in Table 11.6 (see also www.faa.gov/arp/topics.htm).

Table 11.4 World airports ranking by total aircraft movements - 1999–2000

Rank	*Airport*	*Total aircraft movements*	*% change over year*
1	Atlanta (ATL)	909 911	7.4
2	Chicago (ORD)	896 228	n.a.
3	Dallas/Ft Worth airport (DFW)	831 959	–0.5
4	Los Angeles (LAX)	764 653	1.2
5	Phoenix (PHX)	562 714	4.6
6	Detroit (DTW)	559 546	3.8
7	Las Vegas (LAS)	542 922	15.3
8	Oakland (OAK)	524 203	3.5
9	Miami (MIA)	519 861	–3.1
10	Minneapolis/ St Paul (MSP)	510 421	5.7
11	St Louis (STL)	502 865	–2
12	Long Beach (LGB)	499 090	5.8
13	Boston (BOS)	494 816	–2.5
14	Denver (DEN)	488 201	5.3
15	Philadelphia (PHL)	480 276	2.3
16	Cincinnati (Hebron) (CVG)	476 128	7.7
17	Paris (CDG)	475 731	10.7
18	Santa Ana (SNA)	471 676	12.9
19	Washington (IAD)	469 086	22.7
20	Houston (IAH)	463 173	3.5
21	London (LHR)	458 270	1.5
22	Newark (EWR)	457 235	0.3
23	Frankfurt/Main (FRA)	439 093	5.5
24	San Francisco (SFO)	438 685	1.5
25	Pittsburgh (PIT)	437 587	–3
26	Seattle (SEA)	434 425	6.6
27	Charlotte (CLT)	432 128	–2.2
28	Toronto (YYZ)	427 315	1
29	Amsterdam (AMS)	409 999	4.4
30	Memphis (MEM)	374 817	

Table 11.5 Ranking by passenger throughput

	Airport	*Passenger throughput*
1	Atlanta (ATL)	78 092 940
2	Chicago (ORD)	72 609 191
3	Los Angeles (LAX)	64 279 571
4	London (LHR)	62 263 365
5	Dallas/Ft Worth airport (DFW)	60 000 127
6	Tokyo (HND)	54 338 212
7	Frankfurt/Main (FRA)	45 838 864
8	Paris (CDG)	43 597 194
9	San Francisco (SFO)	40 387 538
10	Denver (DEN)	38 034 017
11	Amsterdam (AMS)	36 772 015
12	Minneapolis/St Paul (MSP)	34 721 879
13	Detroit (DTW)	34 038 381
14	Miami (MIA)	33 899 332
15	Las Vegas (LAS)	33 669 185
16	Newark (EWR)	33 622 686
17	Phoenix (PHX)	33 554 407
18	Seoul (SEL)	33 371 074
19	Houston (IAH)	33 051 248
20	New York (JFK)	31 700 604
21	London (LGW)	30 559 227
22	St Louis (STL)	30 188 973
23	Hong Kong (HKG)	29 728 145
24	Orlando (MCO)	29 203 755
25	Madrid (MAD)	27 994 193
26	Toronto (YYZ)	27 779 675
27	Seattle (SEA)	27 705 488
28	Bangkok (BKK)	27 289 299
29	Boston (BOS)	27 052 078
30	Singapore (SIN)	26 064 645

Source of data: ACI.

Table 11.6 FAA–AAS airport related documents

- Airport Ground Vehicle Operations Guide
- Airports (150 Series) Advisory Circulars
- Airports (150 Series) Advisory Circulars (Draft)
- 5010 Data (Airport Master Record) AAS-300
- Access for Passengers With Disabilities
- Activity Data
- AIP APP-500
- AIP Advisory Circular List
- AIP Grants Lists APP-520
- AIP Project Lists APP-520
- Aircraft Rescue and Firefighting Criteria AAS-100
- AC 150/5210-13A Water Rescue Plans, Facilities, and Equipment
- AC 150/5210-14A Airport Fire and Rescue Personnel Protective Clothing
- AC 150/5210-17 Programs for Training of Aircraft Rescue and Firefighting Personnel
- AC 150/5210-18 Systems for Interactive Training of Airport Personnel
- AC 150/5210-19 Driver's Enhanced Vision System (DEVS)
- AC 150/5220-4B Water Supply Systems for Aircraft Fire and Rescue Protection
- AC 150/5220-10B Guide Specification for Water Foam Aircraft Rescue and Firefighting Vehicles
- AC 150/5220-19 Guide Specification for Small Agent Aircraft Rescue and Firefighting Vehicles
- Aircraft Rescue and Firefighting Regulations AAS-310
- Aircraft/Wildlife Strikes (Electronic Filing) (AAS-310)
- Airport Activity Data
- Airport Buildings Specifications AAS-100
- AC 150/5220-18 Buildings for Storage and Maintenance of Airport Snow and Ice Control Equipment and Materials
- Airport Capacity and Delay AAS-100
- Airport Capital Improvement Plan (ACIP)
- Airport Certification (FAR Part 139) AAS-310
- Airport Construction Equipment/Materials Specifications AAS-200
- Airport Construction Specifications AAS-200
- AC 150/5370-10A Standards for Specifying Construction of Airports (includes changes 1–8)
- Airport Design/Geometry AAS-100
- AC 150/5300-13 Airport Design
- Airport Environmental Handbook (FAA Order 5050.4A) APP-600
- Airport Financial Assistance APP-500
- Airport Financial Reports
- Airport Grants APP-500
- Airport Improvement Program (AIP) APP-500

Table 11.6 ***Continued***

- Airport Improvement Program Advisory Circular List
- Airport Lighting AAS-200
- AC 150/5000-13 Announcement of Availability: RTCA Inc., Document RTCA-221
- AC 150/5340-26 Maintenance of Airport Visual Aid Facilities
- AC 150/5345-43E Specification for Obstruction Lighting Equipment
- AC 150/5345-44F Specification for Taxiway and Runway Signs
- AC 150/5345-53B Airport Lighting Equipment Certification Program Addendum
- Airport Lists AAS-330
- Airport Marking AAS-200
- Airport Noise Compatibility Planning (Part 150) APP-600
- Airport Operations Criteria AAS-100
- Airport Operations Equipment Specifications AAS-100
- AC 150/5210-19 Driver's Enhanced Vision System (DEVS)
- AC 150/5220-4B Water Supply Systems for Aircraft Fire and Rescue Protection
- AC 150/5220-10A Guide Specification for Water/Foam Aircraft Rescue and Firefighting Vehicles
- AC 150/5220-19 Guide Specification for Small Agent Aircraft Rescue and Firefighting Vehicles
- AC 150/5220-21A Guide Specification for Lifts Used to Board Airline Passengers with Mobility Impairments
- AC 150/5300-14 Design of Aircraft De-icing Facilities
- Airport Pavement Design AAS-200
- AC 150/5320-16 Airport Pavement Design for the Boeing 777 Airplane
- Airport Planning APP-400
- Airport Privatization (AAS-400)
- Airport Safety & Compliance AAS-400
- Airport Safety Data (Airport Master Record) AAS-330
- Airport Signs, Lighting and Marking AAS-200
- AC 150/5000-13 Announcement of Availability: RTCA Inc., Document RTCA-221
- AC 150/5340-26 Maintenance of Airport Visual Aid Facilities
- AC 150/5345-43E Specification for Obstruction Lighting Equipment
- AC 150/5345-44F Specification for Taxiway and Runway Signs
- AC 150/5345-53A Airport Lighting Equipment Certification Program

Table 11.6 *Continued*

- Airport Statistics
- Airport Visual Aids AAS-200
- AC 150/5000-13 Announcement of Availability: RTCA Inc., Document RTCA-221
- AC 150/5340-26 Maintenance of Airport Visual Aid Facilities
- AC 150/5345-43E Specification for Obstruction Lighting Equipment
- AC 150/5345-44F Specification for Taxiway and Runway Signs
- AC 150/5345-53B Airport Lighting Equipment Certification Program Addendum
- Airports Computer Software
- Airport Planning & Development Process
- Airports Regional/District/Field Offices
- Anniversary
- Announcements
- ARFF Criteria AAS-100
- ARFF Regulations AAS-310
- Aviation State Block Grant Program APP-510
- Benefit and Cost Analysis (APP-500)
- Bird Hazards AAS-310
- AC 150/5200-33, Hazardous Wildlife Attractants on or Near Airports
- Bird Strike Report
- Bird Strikes (Electronic Filing) (AAS-310)
- Bird Strikes (More Information) (AAS-310)
- Buildings Specifications AAS-100
- Capacity and Delay AAS-100
- CertAlerts
- 5010 Data (Airport Master Record) AAS-330
- Certification (FAR Part 139) AAS-310
- Compliance AAS-400
- Compressed Files
- Computer Software
- Construction Equipment/Materials Specifications AAS-200
- Construction Specifications AAS-200
- Declared Distances
- Disabilities
- District/Field Offices
- Draft Advisory Circulars
- Electronic Bulletin Board System
- Emergency Operations Criteria AAS-100
- Emergency Operations Regulations AAS-310
- Engineering Briefs
- Environmental Handbook (FAA Order 5050.4A) APP-600
- Environmental Needs APP-600
- FAA Airport Planning & Development Process

Table 11.6 ***Continued***

- FAA Airports Regional/District/Field Offices
- FAA Airport Safety Newsletter
- FAR Part 139 AAS-310
- FAR Part 150 APP-600
- FAR Part 161 APP-600
- FAR Index
- Federal Register Notices
- Field Offices
- Financial Assistance APP-500
- Financial Reports
- Foreign Object Debris/Damage (FOD) AAS-100
- AC 150/5380-5B Debris Hazards at Civil Airports
- Friction/Traction
- AC 150/5320-12C Measurement, Construction, and Maintenance of Skid-Resistant Airport Pavement Surfaces
- AC 150/5200-30A Airport Winter Safety and Operations
- Fuel Handling and Storage AAS-310
- Grants APP-500
- Grant Assurances APP-510
- Heliport Design AAS-100
- AC 150/5390-2A Heliport Design
- Land Acquisition and Relocation Assistance APP-600
- Legal Notices
- Lighting AAS-200
- AC 150/5000-13 Announcement of Availability: RTCA Inc., Document RTCA-221
- AC 150/5340-26 Maintenance of Airport Visual Aid Facilities
- AC 150/5345-43E Specification for Obstruction Lighting Equipment
- AC 150/5345-44F Specification for Taxiway and Runway Signs
- AC 150/5345-53A Airport Lighting Equipment Certification Program Addendum
- Lighting Equipment Certification Program
- AC 150/5345-53A Airport Lighting Equipment Certification Program Addendum
- List of Advisory Circulars for AIP Projects
- List of Advisory Circulars for PFC Projects
- Marking AAS-200
- Materials Specifications AAS-200
- Military Airport Program (MAP)
- National Plan of Integrated Airports (NPIAS)
- National Priority System
- Newsletter – FAA Airport Safety Newsletter
- Noise Compatibility Planning (Part 150) APP-600
- Notice and Approval of Airport Noise and Access Restrictions (Part 161) APP-600

Table 11.6 *Continued*

- Notices
- Notices to Airmen (NOTAMs) AAS-310
- AC 150/5200-28B, Notices to Airmen (NOTAMs) for Airport Operators
- Obstruction Lighting AAS-200
- Operations Criteria AAS-100
- Operations Equipment Specifications AAS-100
- Part 139 AAS-310
- Part 150 APP-600
- Part 161 APP-600
- Passenger Facility Charges (PFC) APP-530
- Passenger Facility Charges Advisory Circular List
- Passengers with Disabilities
- Pavement Design AAS-200
- PFC APP-530
- PFC Advisory Circular List
- Planning APP-400
- Privatization AAS-400
- Radio Control Equipment AAS-200
- Regional/Field Offices
- Relocation Assistance APP-600
- Runway Friction/Traction
- Runway Guard Lights
- AC 150/5000-13 Announcement of Availability: RTCA Inc., Document RTCA-221
- Safety & Compliance AAS-400
- Safety Data (Airport Master Record) AAS-330
- Safety Newsletter – FAA Airport Safety Newsletter
- Seaplane Bases AAS-100
- AC 150/5395-1 Seaplane Bases
- Signs, Lighting and Marking AAS-200
- Signs and Marking Supplement (SAMS)
- Snow/Ice AAS-100
- Statistics
- Strikes: Bird/Wildlife (Electronic Filing) (AAS-310)
- Surface Movement Guidance and Control Systems (SMGCS)
- Traction
- Training – FY 2000 Airports Training Class Schedule
- Vertiport Design AAS-100
- Visual Aids AAS-200
- Wildlife Control AAS-310
- AC 150/5200-33, Hazardous Wildlife Attractants on or Near Airports
- Bird Strike Report
- Wildlife Strikes (Electronic Filing) (AAS-310)
- Wildlife Strikes (More Information) (AAS-310)
- Winter Operations Criteria AAS-100
- Winter Operations Regulations AAS-310

11.5 Worldwide airport geographical data

Table 11.7 gives details of the geographical location of major world civil airports

11.6 Airport reference sources and bibliography

1. Norman Ashford and Paul H. Wright, *Airport Engineering*, 3rd ed. (1992), comprehensively sets forth the planning, layout, and design of passenger and freight airports, including heliports and short take-off and landing (STOL) facilities.
2. Robert Horonjeff and Francis X. McKelvey, *Planning and Design of Airports*, 4th ed. (1993), is a comprehensive civil engineering text on the planning, layout, and design of airports with strong emphasis on aspects such as aircraft pavements and drainage.
3. International Civil Aviation Organization, *Aerodromes: International Standards and Recommended Practices* (1990), includes the internationally adopted design and operational standards for all airports engaged in international civil aviation.
4. Christopher R. Blow, *Airport Terminals* (1991), provides an architectural view of the functioning of airport passenger terminals with extensive coverage of design case studies. Walter Hart, *The Airport Passenger Terminal* (1985, reprinted 1991), describes the functions of passenger terminals and their design requirements.
5. International Air Transport Association, *Airport Terminals Reference Manual*, 7th ed. (1989), provides design and performance requirements of passenger and freight terminals as set out by the international airlines' trade association.
6. Denis Phipps, *The Management of Aviation Security* (1991), describes the operational and design requirements of civil airports to conform to national and international regulations.
7. Norman Ashford, H.P. Martin Stanton, and Clifton A. Moore, *Airport Operations* (1984, reissued 1991), extensively discusses many aspects of airport operation and management, including administrative structure, security, safety, environmental impact, performance indices, and passenger and aircraft handling.
8. Norman Ashford and Clifton A. Moore, *Airport Finance* (1992), discusses the revenue and expenditure patterns of airport authorities, methods of financing, business planning, and project appraisal.
9. Rigas Doganis, *The Airport Business* (1992), examines the status of airport business in the early 1990s, performance indices, commercial opportunities, and privatization of airports.

Table 11.7 Worldwide airport data

City name	*Airport name*	*Country*	*Length (ft)*	*Elevation (ft)*	*Geographic location*
Anchorage Intl	Anchorage Intl	Alaska	10 897	144	6110N 15000W
Fairbanks	Fairbanks Intl	Alaska	10 300	434	6449N 14751W
Buenos Aires	Ezeiza	Argentina	10 827	66	3449S 5832W
Ascension	Wideawake	Ascension Is.	104 000	273	0758S 1424W
Alice Springs	Alice Springs	Australia	8000	1789	2349S 13354E
Brisbane	Brisbane	Australia	11 483	13	2723S 15307E
Cairns	Cairns	Australia	10 489	10	1653S 1454E
Canberra	Canberra	Australia	8800	1888	3519S 14912E
Darwin	Darwin Intl	Australia	10 906	102	1225S 13053E
Melbourne	Melbourne Intl	Australia	12 000	434	3741S 14451E
Sydney	Kingford Smith	Australia	13 000	21	3357S 15110E
Innsbruck	Innsbruck	Austria	6562	1906	4716N 1121E
Salzburg	Salzburg	Austria	8366	1411	4748N 1300E
Vienna	Schwechat	Austria	11 811	600	4807N 1633E
Baku	Bina	Azerbaijan	8858	0	4029N 5004E
Freeport	Freeport	Bahamas	11 000	7	2633N 7842W
Bahrain	Bahrain Intl	Bahrain	13 002	6	2616N 5038E
Chittagong	Chittagong	Bangladesh	10 000	12	2215N 9150E

Barbados	Grantly Adams Intl	Barbados	11 000	169	1304N 5930W
Minsk	Minsk-2	Belarus	11 942	669	5353N 2801E
Antwerp	Deurne	Belgium	4839	39	5111N 0428E
Brussels	Brussels National	Belgium	11 936	184	5054N 0429E
Brasilia	Brasilia	Brazil	10 496	3474	1551S 4754W
Rio De Janeiro	Galeao Intl	Brazil	13 123	30	2249S 4315W
São Paulo	Guarulhas	Brazil	12 140	2459	2326S 4629W
Ouagadougou	Ouagadougou	Burkina	9842	1037	1221N 0131W
Douala	Douala	Cameroon	9350	33	0401N 0943E
Halifax	Halifax Intl	Canada	8800	476	4453N 6331S
Quebec	Quebec	Canada	9000	243	4648N 7123W
Toronto	Toronto	Canada	11 050	569	4341N 7938W
Vancouver	Vancouver	Canada	11 000	9	4911N 12310W
Yellowknife	Yellowknife	Canada	7500	675	6228N 11427W
Gran Canaria	Las Palmas	Canary Is.	10 170	75	2756N 1523W
Lanzarote	Lanzarote	Canary Is.	7874	46	2856N 1336W
Beijing	Capital	China	12 467	115	4004N 11635E
Chengdu	Shuangliu	China	9186	1624	3035N 10357E
Shanghai	Hongqiac	China	10 499	10	3112N 12120E
Urumqi	Diwopu	China	10 499	2129	4354N 8729E
Bogota	Eldorado	Colombia	12 467	8355	0442N 7409W
Zagreb	Zagreb	Croatia	10 663	351	4545N 1604E

Table 11.7 Worldwide airport data – *Continued*

City name	*Airport name*	*Country*	*Length (ft)*	*Elevation (ft)*	*Geographic location*
Havana	Jose Marti Intl	Cuba	13 123	210	2300N 8225W
Paphos	Paphos Intl	Cyprus	8858	41	3443N 3229E
Prague	Ruzyne	Czech Republic	12 188	1247	5006N 1416E
Copenhagen Kastrup	Kastrup	Denmark	11 811	17	5537N 1239E
Cairo	Cairo Intl	Egypt	10 827	381	3007N 3124E
Helsinki Malmi	Malmi	Finland	4590	57	6051N 2503E
Basle	Mulhouse	France	12 795	883	4735N 0732E
Lyon	Bron	France	5971	659	4544N 0456E
Paris Charles De Gaulle	Charles-De-Gaulle	France	11 860	387	4901N 0233E
Paris Orly	Orly	France	11 975	292	4843N 0223E
Strasbourg	Entzheim	France	7874	502	4832N 0738E
Tarbes	Ossun–Lourdes	France	9843	1243	4311N 0000E
Berlin Tegel	Tegel	Germany	9918	121	5234N 1317E
Cologne–Bonn	Cologne–Bonn	Germany	12 467	300	5052N 0709E
Düsseldorf	Düsseldorf	Germany	9843	147	5117N 0645E
Frankfurt	Main	Germany	13 123	365	5002N 0834E
Hamburg	Hamburg	Germany	12 028	53	5338N 0959E
Leipzig	Halle	Germany	8202	466	5125N 1214E

Munich	Munich	Germany	13 123	1486	4821N 1147E
Stuttgart	Stuttgart	Germany	8366	1300	4841N 0913e
Takoradi	Takoradi	Ghana	5745	21	0454N 0146W
Gibraltar	Gibraltar	Gibraltar	6000	15	3609N 0521W
Athens	Central	Greece	11 483	68	3754N 2344E
Guatemala	La Aurora	Guatemala	9800	4952	1435N 9032W
Hong Kong	Kai Tak	Hong Kong	11 130	15	2219N 11412E
Budapest	Ferihegy	Hungary	12 162	495	4726N 1916E
Keflavik	Keflavik	Iceland	10 013	171	6359N 2237W
Bombay	Jawaharial Nehru Intl	INDIA	11 447	26	1905N 7252E
Calcutta	NS Chandra Bose Intl	India	11 900	18	2239N 8827E
Delhi	Delhi Intl	India	12 500	744	2834N 7707E
Bali	Bali Intl	Indonesia	9843	14	0845S 11510E
Jakarta Intl	Soerkarno-Hatta Intl	Indonesia	12 008	34	0608S 10639E
Tehran	Mehrabad	Iran	13 123	3962	3541N 5119E
Cork	Cork	Ireland	7000	502	5150N 0829W
Dublin	Dublin	Ireland	8652	242	5326N 0615W
Shannon	Shannon	Ireland	10 500	47	5242N 0855W
Tel Aviv	Ben Gurion Intl	Israel	11 998	135	3201N 3453E
Milan Malpensa	Malpensa	Italy	12 844	767	4538N 0843E
Naples	Naples	Italy	8661	296	4053N 1417E
Pisa	Pisa	Italy	9800	9	4341N 1024E

Table 11.7 Worldwide airport data – *Continued*

City name	*Airport name*	*Country*	*Length (ft)*	*Elevation (ft)*	*Geographic location*
Kingston	Kingston	Jamaica	8786	10	1756N 7648W
Montego Bay	Sangster Intl	Jamaica	8705	4	1830N 7755W
Nagasaki	Nagasaki	Japan	9840	8	3255N 12955E
Tokyo Narita	Narita	Japan	13 123	135	3546N 14023E
Mombasa	Moi	Kenya	10 991	196	0402S 3936E
Nairobi	Jomo Kenyatta	Kenya	13 507	5327	0119S 3656E
Tripoli	Tripoli Intl	Libya	11 811	263	3240N 1309E
Tombouctou	Tombouctou	Mali	4921	863	1644N 0300W
Acapulco	Acapulco Intl	Mexico	10 824	16	1645N 9945W
Cancun	Cancun	Mexico	11 484	23	2102N 8653W
Mexico City	B. Juarez Intl	Mexico	12 795	7341	3193N 9904W
Kathmandu	Tribhuvan	Nepal	10 007	4390	2742S 8522E
Amsterdam	Schipol	Netherlands	11 330	–11	5218N 0446E
Rotterdam	Rotterdam	Netherlands	7218	–14	5157N 0426E
Auckland	Auckland Intl	New Zealand	11 926	23	3701S 17447E
Wellington	Wellington Intl	New Zealand	6350	40	4120S 17448E
Lagos	Murtala Muhammed	Nigeria	12 795	135	0635N 0319E
Bergen	Flesland	Norway	8038	165	6018N 0513E

Stavanger	Sola	Norway	8383	29	5853N 0538E
Tromsö	Tromsö	Norway	7080	29	6941N 1855E
Muscat	Seeb	Oman	11 762	48	2336N 5817E
Karachi	Karachi	Pakistan	10 500	100	2454N 6709E
Warsaw	Okecie	Poland	12 106	361	5210N 2058E
Faro	Faro	Portugal	8169	24	3701N 0758W
San Juan	Luis Munoz Marin Intl	Puerto Rico	10 000	10	1826N 6600W
Doha	Doha	Qatar	15 000	35	2516N 5134E
Bucharest Baneasa	Baneasa	Romania	9843	295	4430N 2606E
Moscow Shremetievo	Sheremetievo	Russia	12 139	627	5558N 3725E
Novosibirsk	Tolmachevo	Russia	11 808	364	5501N 8240E
St Petersburg	Pulkovo	Russia	12 408	79	5948N 3016E
Dharan	Dharan	Saudi Arabia	12 008	84	2617N 5010E
Jeddah	King Abdulaziz	Saudi Arabia	12 467	48	2141N 3909E
Riyadh	King Khalid Intl	Saudi Arabia	13 780	2049	2458N 4643E
Dakar	Yoff	Senegal	11 450	89	1445N 1730W
Seychelles	Seychelles Intl	Seychelles	9800	10	0440S 5531E
Singapore Changi	Changi	Singapore	13 123	23	0122N 10359E
Mogadishu	Mogadishu	Somalia Republic	10 335	27	0200N 4518E
Cape Town	D.F. Malan	South Africa	10 500	151	3358S 1836E
Durban Virginia	Virginia	South Africa	3051	20	2946S 3104E
Johannesburg Intl	Jan Smuts	South Africa	14 495	5557	2608S 2815E

Table 11.7 Worldwide airport data – *Continued*

City name	*Airport name*	*Country*	*Length (ft)*	*Elevation (ft)*	*Geographic location*
Pretoria	Wonderbroom	South Africa	6000	4095	2539S 2813E
Seoul	Kimpo Intl	South Korea	11 811	58	3733N 12648E
Barcelona	Barcelona	Spain	10 197	13	4118N 0205W
Madrid Barajas	Barajas	Spain	13 450	1999	4029N 0334W
Palma	Palma	Spain	10 728	32	3933N 0244E
Valencia	Valencia	Spain	8858	226	3929N 0029W
Khartoum	Khartoum	Sudan	9843	1261	1535N 3233E
Malmo	Sturup	Sweden	9186	236	5533N 1322E
Stockholm Arlanda	Arlanda	Sweden	10 827	123	5939N 1755E
Zürich	Zürich	Switzerland	12 140	1416	4728N 0833E
Damascus	Damascus Intl	Syria	11 811	2020	3325N 3631E
Taipei Intl	Chiang Kai Shek	Taiwan	12 008	73	2505N 12113E
Bangkok	Bangkok	Thailand	12 139	9	1355N 10037E
Istanbul	Ataturk	Turkey	9842	158	4059N 2849E
Entebbe	Entebbe	Uganda	12 001	3782	0003N 3226E
Abu Dhabi	Abu Dhabi Intl	United Arab Emirates	13 451	88	2426N 5439E
Dubai	Dubai	United Arab Emirates	13 123	34	2515N 5521E
Belfast	City	United Kingdom	6000	15	5437N 0552W

Birmingham UK	Birmingham	United Kingdom	7398	325	5227N 0145W
Bristol	Bristol	United Kingdom	6598	620	5123N 0243W
Cardiff	Cardiff	United Kingdom	7000	220	5124N 0321W
East Midlands	East Midlands	United Kingdom	7480	310	5250N 0119W
Glasgow	Glasgow	United Kingdom	8720	26	5552N 0426W
Leeds Bradford	Leeds Bradford	United Kingdom	7382	681	5352N 0140W
London City	City	United Kingdom	3379	16	5130N 0003E
London Gatwick	Gatwick	United Kingdom	10 364	202	5109N 0011W
London Heathrow	Heathrow	United Kingdom	12 802	80	5129N 0028W
London Stansted	Stansted	United Kingdom	10 000	347	5153N 0014E
Luton	Luton	United Kingdom	7087	526	5153N 0022W
Manchester	Manchester	United Kingdom	10 000	256	5321N 0216W
Newcastle	Newcastle	United Kingdom	7651	266	5502N 0141W
Atlanta	Wm. B. Hartsfield	United States	11 889	1026	3338N 8426W
Baltimore	Washington Intl	United States	9519	146	3911N 7640W
Boston	Logan Intl	United States	10 081	20	4222N 7100W
Chicago	Chicago O'hare	United States	13 000	667	4159N 8754W
Cincinnati	Northern Kentucky Intl	United States	10 000	891	3903N 8440W
Denver	Denver Intl	United States	12 000	5431	3951N 10440W
Des Moines	Des Moines	United States	9000	957	4132N 9339W
Houston	Houston Intl	United States	12 000	98	2959N 9520W
Las Vegas	Las Vegas	United States	12 635	2174	3605N 11509W

Table 11.7 Worldwide airport data – *Continued*

City name	*Airport name*	*Country*	*Length (ft)*	*Elevation (ft)*	*Geographic location*
Los Angeles	Los Angeles Intl	United States	12 090	126	3356N 11824W
Miami	Miami Intl	United States	13 000	10	2548N 8017W
New York John F. Kennedy	John F. Kennedy	United States	14 572	12	4039N 7374W
Philadelphia	Philadelphia	United States	10 500	21	3953N 7514W
Pittsburgh	Pittsburgh	United States	11 500	1203	4030N 8014W
Salt Lake City	Salt Lake City	United States	12 000	4227	4047N 11158W
San Diego	San Diego	United States	9400	15	3244N 11711W
San Francisco	San Francisco	United States	11 870	11	3737N 12223W
Seattle	Tacoma	United States	11 900	429	4727N 12218W
Washington Dulles	Dulles	United States	11 500	313	3857N 7727W
Tashkent	Yuzhnyy	Uzbekistan	13 123	1414	4115N 6917E
Caracas	Simon Bolivar	Venezuela	11 483	235	1036N 6659W
Hanoi	Noibai	Vietnam	10 499	39	2113N 10548E
Belgrade	Belgrade	Yugoslavia	11 155	335	4449N 2019E
Kinshasa	Ndjili	Zaire	11 811	1027	0423S 1526E
Harare	Charles Prince	Zimbabwae	3035	4850	1745S 3055E

Section 12

Basic mechanical design

The techniques of basic mechanical design are found in all aspects of aeronautical engineering.

12.1 Engineering abbreviations

The following abbreviations, based on the published standard ANSI/ASME Y14.5 81: 1994: *Dimensioning and Tolerancing*, are in common use in engineering drawings and specifications in the USA (Table 12.1).

In Europe, a slightly different set of abbreviations is used (see Table 12.2).

12.2 Preferred numbers and preferred sizes

Preferred numbers are derived from geometric series, in which each term is a uniform percentage larger than its predecessor. The first five principal series (named the 'R' series) are shown in Figure 12.1. Preferred numbers are taken as the basis for ranges of linear sizes of components, often being rounded up or down for convenience. Figure 12.2 shows the development of the R5 and R10 series.

Series	Basis	Ratio of terms (% increase)
R5	$5\sqrt{10}$	1.58 (58%)
R10	$10\sqrt{10}$	1.26 (26%)
R20	$20\sqrt{10}$	1.12 (12%)
R40	$40\sqrt{10}$	1.06 (6%)
R80	$80\sqrt{10}$	1.03 (3%)

Fig. 12.1 The first five principal 'R' series

Table 12.1 Engineering abbreviations: USA

Abbreviation	*Meaning*
ANSI	American National Standards Institute
ASA	American Standards Association
ASME	American Society of Mechanical Engineers
AVG	average
CBORE	counterbore
CDRILL	counterdrill
CL	center line
CSK	countersink
FIM	full indicator movement
FIR	full indicator reading
GD&T	geometric dimensioning and tolerancing
ISO	International Standards Organization
LMC	least material condition
MAX	maximum
MDD	master dimension definition
MDS	master dimension surface
MIN	minimum
mm	millimeter
MMC	maximum material condition
PORM	plus or minus
R	radius
REF	reference
REQD	required
RFS	regardless of feature size
SEP REQT	separate requirement
SI	Système International (the metric system)
SR	spherical radius
SURF	surface
THRU	through
TIR	total indicator reading
TOL	tolerance

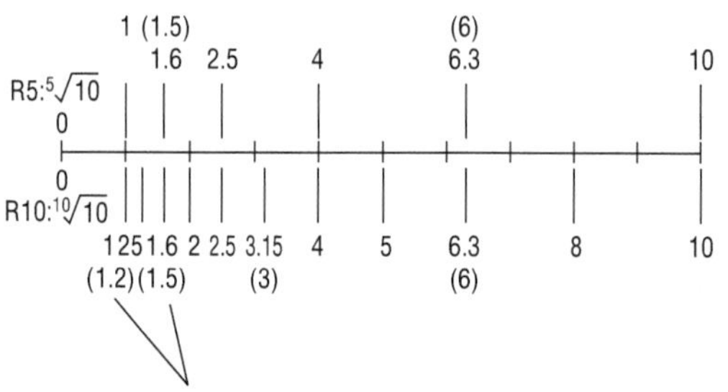

Fig. 12.2 The R5 and R10 series

Table 12.2 Engineering abbreviations in common use: Europe

Abbreviation	*Meaning*
A/F	Across flats
ASSY	Assembly
CRS	Centres
L or CL	Centre line
CHAM	Chamfered
CSK	Countersunk
C'BORE	Counterbore
CYL	Cylinder or cylindrical
DIA	Diameter (in a note)
∅	Diameter (preceding a dimension)
DRG	Drawing
EXT	External
FIG.	Figure
HEX	Hexagon
INT	Internal
LH	Left hand
LG	Long
MATL	Material
MAX	Maximum
MIN	Minimum
NO.	Number
PATT NO.	Pattern number
PCD	Pitch circle diameter
RAD	Radius (in a note)
R	Radius (preceding a dimension)
REQD	Required
RH	Right hand
SCR	Screwed
SH	Sheet
SK	Sketch
SPEC	Specification
SQ	Square (in a note)
□	Square (preceding a dimension)
STD	Standard
VOL	Volume
WT	Weight

12.3 Datums and tolerances – principles

A *datum* is a reference point or surface from which all other dimensions of a component are taken; these other dimensions are said to be *referred to* the datum. In most practical designs, a datum surface is normally used, this generally being one of the surfaces of the machine element

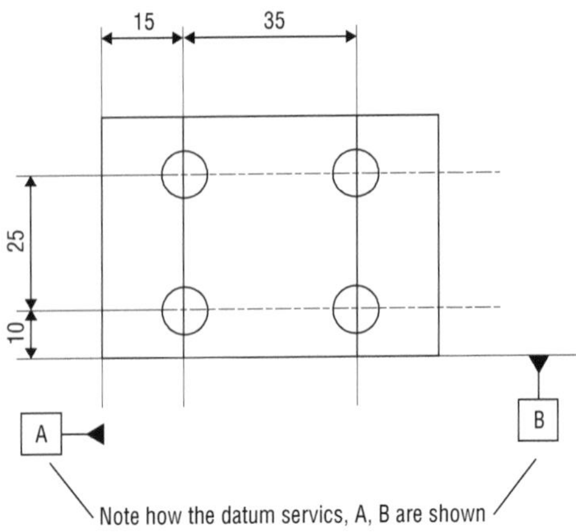

Fig. 12.3 Datum surfaces

itself rather than an 'imaginary' surface. This means that the datum surface normally plays some important part in the operation of the elements – it is usually machined and may be a mating surface or a locating face between elements, or similar (see Figure 12.3). Simple machine mechanisms do not *always* need datums; it depends on what the elements do and how complicated the mechanism assembly is.

A *tolerance* is the allowable variation of a linear or angular dimension about its 'perfect' value. British Standard BS 308: 1994 contains accepted methods and symbols (see Figure 12.4).

12.4 Toleranced dimensions

In designing any engineering component it is necessary to decide which dimensions will be toleranced. This is predominantly an exercise in necessity – only those dimensions that *must* be tightly controlled, to preserve the functionality of the component, should be toleranced. Too many toleranced dimensions will increase significantly the manufacturing costs and may result in 'tolerance clash', where a dimension derived from other toleranced dimensions

BS 308

	Tolerance characteristic
—	Straightness
▱	Flatness
○	Roundness
//	Parallelism
∠	Angularity
⊥	Squareness
◎	Concentricity
↗	Run-out
⌰	Total run-out

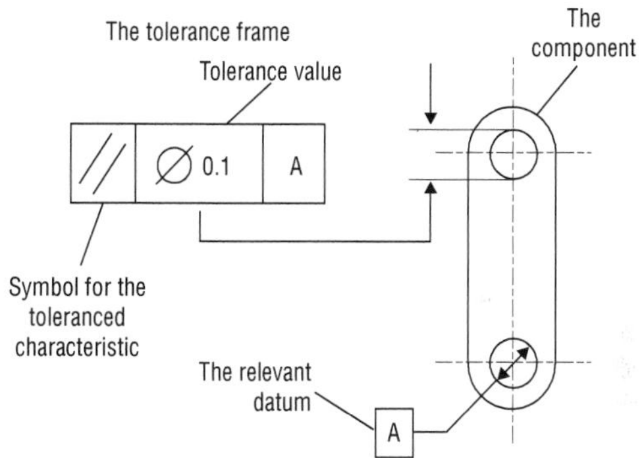

Fig. 12.4 Tolerancing symbols

can have several contradictory values (see Figure 12.5).

12.4.1 General tolerances

It is a sound principle of engineering practice that in any machine design there will only be a small number of toleranced features. The remainder of the dimensions will not be critical. There are two ways to deal with this: first, an engineering drawing or sketch can be

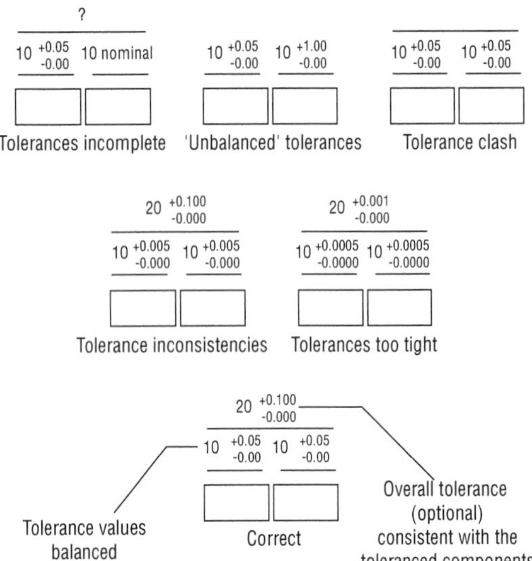

Fig. 12.5 Toleranced dimensions

annotated to specify that a *general tolerance* should apply to features where no specific tolerance is mentioned. This is often expressed as ±0.020 in or '20 mils' (0.5 mm).

12.4.2 Holes

The tolerancing of holes depends on whether they are made in thin sheet (up to about 1/8 in (3.2 mm) thick) or in thicker plate material. In thin material, only two toleranced dimensions are required:

- *Size*: A toleranced diameter of the hole, showing the maximum and minimum allowable dimensions.
- *Position*: Position can be located with reference to a datum and/or its spacing from an adjacent hole. Holes are generally spaced by reference to their centres.

For thicker material, three further toleranced dimensions become relevant: straightness, parallelism and squareness (see Figure 12.6).

Straightness

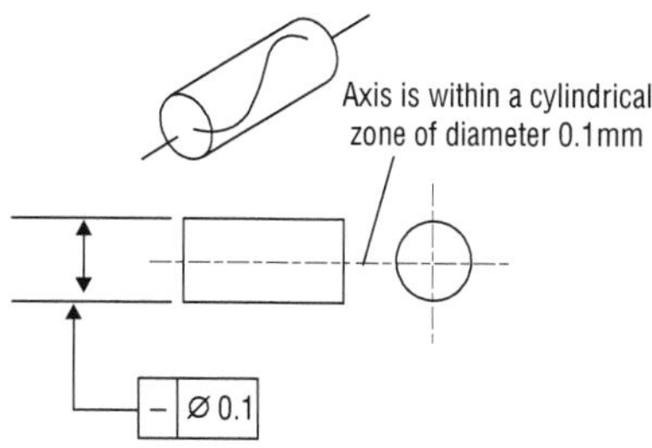

Squareness

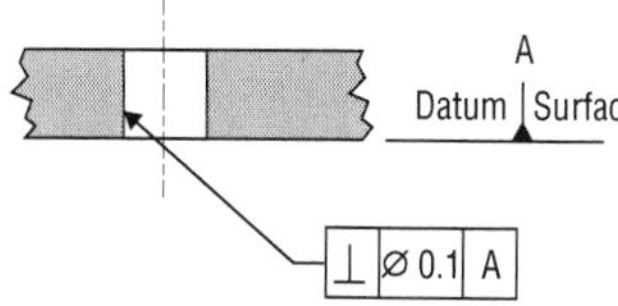

Axis of hole to be within a cylindrical zone of diameter 0.1mm at 90° to the datum surface A

Parallelism

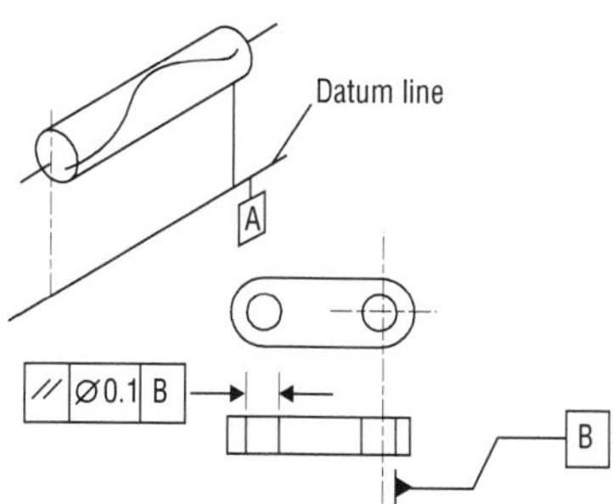

Axis is within a cylindrical zone of diameter 0.1mm parallel to the datum line A

Fig. 12.6 Straightness, parallelism and squareness

- *Straightness*: A hole or shaft can be *straight* without being perpendicular to the surface of the material.
- *Parallelism*: This is particularly relevant to holes and is important when there is a mating hole-to-shaft fit.

- *Squareness*: The formal term for this is perpendicularity. Simplistically, it refers to the squareness of the axis of a hole to the datum surface of the material through which the hole is made.

12.4.3 Screw threads

There is a well-established system of tolerancing adopted by ANSI/ASME, International Standard Organizations and manufacturing industry. This system uses the two complementary elements of fundamental deviation and tolerance range to define fully the tolerance of a single component. It can be applied easily to components, such as screw threads, which join or mate together (see Figure 12.7).

For screw threads, the tolerance layout shown applies to major, pitch, and minor diameters (although the actual diameters differ).

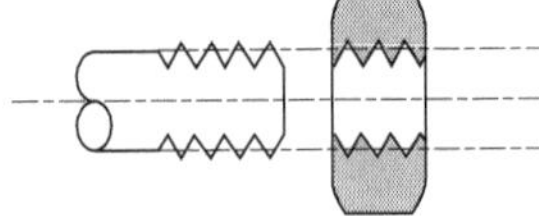

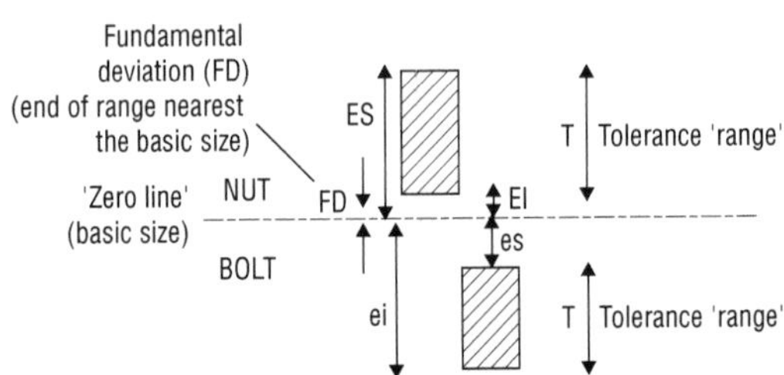

FD is designated by a letter code, e.g. g,H
Tolerance range (T) is designated by a number code, e.g. 5, 6, 7

Commonly used symbols are:
EI – lower deviation (nut)
ES – upper deviation (nut)
ei – lower deviation (bolt)
es – upper deviation (bolt)

Fig. 12.7 Tolerancing: screw threads

- *Fundamental deviation*: (FD) is the distance (or 'deviation') of the nearest 'end' of the tolerance band from the nominal or 'basic' size of a dimension.
- *Tolerance band*: (or 'range') is the size of the tolerance band, i.e. the difference between the maximum and minimum acceptable size of a toleranced dimension. The size of the tolerance band, and the location of the FD, governs the system of limits and fits applied to mating parts.

Tolerance values have a key influence on the costs of a manufactured item so their choice must be seen in terms of economics as well as engineering practicality. Mass-produced items are competitive and price sensitive, and over-tolerancing can affect the economics of a product range.

12.5 Limits and fits

12.5.1 Principles

In machine element design there is a variety of different ways in which a shaft and hole are required to fit together. Elements such as bearings, location pins, pegs, spindles and axles are typical examples. The shaft may be required to be a tight fit in the hole, or to be looser, giving a clearance to allow easy removal or rotation. The system designed to establish a series of useful fits between shafts and holes is termed *limits and fits*. This involves a series of tolerance grades so that machine elements can be made with the correct degree of accuracy and be interchangeable with others of the same tolerance grade. The standards ANSI B4.1/B4.3 contain the recommended tolerances for a wide range of engineering requirements. Each fit is designated by a combination of letters and numbers (see Tables 12.3, 12.4 and 12.5).

Figure 12.8 shows the principles of a shaft/hole fit. The 'zero line' indicates the basic or 'nominal' size of the hole and shaft (it is the

Table 12.3 Classes of fit (imperial)

1. *Loose running fit*: Class RC8 and RC9. These are used for loose 'commercial-grade' components where a significant clearance is necessary.
2. *Free running fit*: Class RC7. Used for loose bearings with large temperature variations.
3. *Medium running fit*: Class RC6 and RC5. Used for bearings with high running speeds.
4. *Close running fit*: Class RC4. Used for medium-speed journal bearings.
5. *Precision running fit*: Class RC3. Used for precision and slow-speed journal bearings.
6. *Sliding fit*: Class RC2. A locational fit in which close-fitting components slide together.
7. *Close sliding fit*: Class RC1. An accurate locational fit in which close-fitting components slide together.
8. *Light drive fit*: Class FN1. A light push fit for long or slender components.
9. *Medium drive fit*: Class FN2. A light shrink-fit suitable for cast-iron components.
10. *Heavy drive fit*: Class FN3. A common shrink-fit for steel sections.
11. *Force fit*: Class FN4 and FN5. Only suitable for high-strength components.

Table 12.4 Force and shrink fits (imperial)

Nominal size range, in	*Class*				
	FN1	*FN2*	*FN3*	*FN4*	*FN5*
0.04–0.12	0.05	0.2		0.3	0.5
	0.5	0.85		0.95	1.3
0.12–0.24	0.1	0.2		0.95	1.3
	0.6	1.0		1.2	1.7
0.24–0.40	0.1	0.4		0.6	0.5
	0.75	1.4		1.6	2.0
0.40–0.56	0.1	0.5		0.7	0.6
	0.8	1.6		1.8	2.3
0.56–0.71	0.2	0.5		0.7	0.8
	0.9	1.6		1.8	2.5
0.71–0.95	0.2	0.6		0.8	1.0
	1.1	1.9		2.1	3.0
0.95–1.19	0.3	0.6	0.8	1.0	1.3
	1.2	1.9	2.1	2.3	3.3
1.19–1.58	0.3	0.8	1.0	1.5	1.4
	1.3	2.4	2.6	3.1	4.0
1.58–1.97	0.4	0.8	1.2	1.8	2.4
	1.4	2.4	2.8	3.4	5.0
1.97–2.56	0.6	0.8	1.3	2.3	3.2
	1.8	2.7	3.2	4.2	6.2
2.56–3.15	0.7	1.0	1.8	2.8	4.2
	1.9	2.9	3.7	4.7	7.2

Limits in 'mils' (0.001 in).

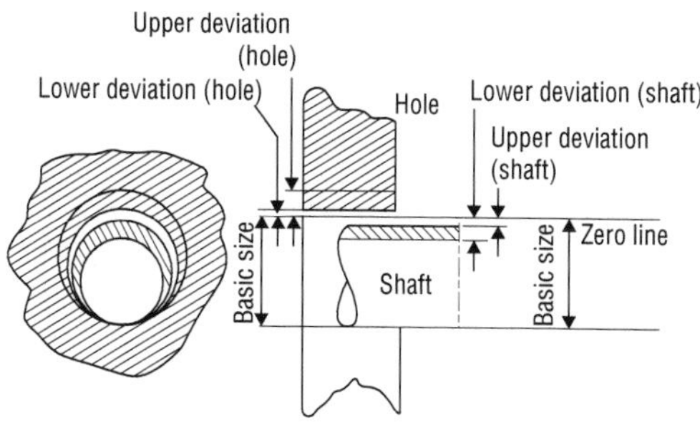

Fig. 12.8 Principles of a shaft–hole fit

Table 12.5 Running and sliding fits (imperial)

Nominal size range, in	*Class*								
	RC1	*RC2*	*RC3*	*RC4*	*RC5*	*RC6*	*RC7*	*RC8*	*RC9*
0–0.12	0.1	0.1	0.3	0.3	0.6	0.6	1.0	2.5	4.0
	0.45	0.55	0.95	1.3	1.6	2.2	2.6	5.1	8.1
0.12–0.24	1.5	0.15	0.4	0.4	0.8	0.8	1.2	2.8	4.5
	0.5	0.65	1.2	1.6	2.0	2.7	3.1	5.8	9.0
0.24–0.40	0.2	0.2	0.5	0.5	1.0	1.0	1.6	3.0	5.0
	0.6	0.85	1.5	2.0	2.5	3.3	3.9	6.6	10.7
0.40–0.71	0.25	0.25	0.6	0.6	1.2	1.2	2.0	3.5	6.0
	0.75	0.95	1.7	2.3	2.9	3.8	4.6	7.9	12.8
0.71–1.19	0.3	0.3	0.8	0.8	1.6	1.6	2.5	4.5	7.0
	0.95	1.2	2.1	2.8	3.6	4.8	5.7	10.0	15.5
1.19–1.97	0.4	0.4	1.0	1.0	2.0	2.0	3.0	5.0	8.0
	1.1	1.4	2.6	3.6	4.6	6.1	7.1	11.5	18.0
1.97–3.15	0.4	0.4	1.2	1.2	2.5	2.5	4.0	6.0	9.0
	1.2	1.6	3.1	4.2	5.5	7.3	8.8	13.5	20.5
3.15–4.73	0.5	0.5	1.4	1.4	3.0	3.0	5.0	7.0	10.0
	1.5	2.0	3.7	5.0	6.6	8.7	10.7	15.5	24.0

Limits in 'mils' (0.001 in).

same for each) and the two shaded areas depict the tolerance zones within which the hole and shaft may vary. The hole is conventionally shown above the zero line. The algebraic difference between the basic size of a shaft or hole and its actual size is known as the *deviation*.

- It is the deviation that determines the nature of the fit between a hole and a shaft.

- If the deviation is small, the tolerance range will be near the basic size, giving a tight fit.
- A large deviation gives a loose fit.

Various grades of deviation are designated by letters, similar to the system of numbers used for the tolerance ranges. Shaft deviations are denoted by small letters and hole deviations by capital letters. Most general engineering uses a 'hole-based' fit in which the larger part of the available tolerance is allocated to the hole (because it is more difficult to make an accurate hole) and then the shaft is made to suit, to achieve the desired fit.

Tables 12.4 and 12.5 show suggested clearance and fit dimensions for various diameters (ref.: ANSI B4.1 and 4.3).

Table 12.6 Metric fit classes

1. *Easy running fit*: H11-c11, H9-d10, H9-e9. These are used for bearings where a significant clearance is necessary.
2. *Close running fit*: H8-f7, H8-g6. This only allows a small clearance, suitable for sliding spigot fits and infrequently used journal bearings. This fit is not suitable for continuously rotating bearings.
3. *Sliding fit*: H7-h6. Normally used as a locational fit in which close-fitting items slide together. It incorporates a very small clearance and can still be freely assembled and disassembled.
4. *Push fit*: H7-k6. This is a transition fit, mid-way between fits that have a guaranteed clearance and those where there is metal interference. It is used where accurate location is required, e.g. dowel and bearing inner-race fixings.
5. *Drive fit*: H7-n6. This is a tighter grade of transition fit than the H7–k6. It gives a tight assembly fit where the hole and shaft may need to be pressed together.
6. *Light press fit*: H7-p6. This is used where a hole and shaft need permanent, accurate assembly. The parts need pressing together but the fit is not so tight that it will overstress the hole bore.
7. *Press fit*: H7-s6. This is the tightest practical fit for machine elements such as bearing bushes. Larger interference fits are possible but are only suitable for large heavy engineering components.

12.5.2 Metric equivalents

The metric system (ref. ISO Standard EN 20286) *ISO 'limits and fits'* uses seven popular combinations with similar definitions (see Table 12.6 and Figure 12.9).

Clearance fits — Holes: H11, H9, H9, H8, H7, H7; Shafts: c11, d10, e9, f7, g6, h6. Transmission fits: H7, H7, k6, p5. Interference fits: p6, H7, s6.

Nominal size in mm	Easy running						Close running				Sliding		Push		Drive		Light press		Press	
	Tols*		Tols		Tols		Tols		Tols		Tols		Tols		Tols		Tols		Tols	
	H11	c11	H9	d10	H9	e9	H8	f7	H7	g6	H7	h6	H7	k6	H7	n6	H7	p6	H7	s6
6-10	+90 0	-80 -170	+36 0	-40 -98	+36 0	-25 -61	+22 0	-12 -28	+15 0	-5 -14	+15 0	-9 0	+15 0	+10 +1	+15 0	+19 +10	+15 0	+24 +15	+15 0	+32 +23
10-18	+110 0	-95 -205	+43 0	-50 -120	+43 0	-32 -75	+27 0	-16 -34	+18 0	-6 -17	+18 0	-11 0	+18 0	+12 +1	+18 0	+23 +12	+18 0	+29 +18	+18 0	+39 +28
18-30	+130 0	-110 -240	+52 0	-69 -149	+52 0	-40 -92	+33 0	-20 -41	+21 0	-7 -20	+21 0	-13 0	+21 0	+15 +2	+21 0	+28 +15	+21 0	+35 +22	+21 0	+48 +35
30-40	+140 0	-120 -280	+62	-80	+62	-50	+39	-25	+25	-9	+25	-16	+25	+18	+25	+33	+25	+42	+25	+59
40-50	+160 0	-130 -290	0	-180	0	-112	0	-50	-50	-25	0	0	0	+2	0	+17	0	+26	0	+43

*Tolerance units in 0.001 mm — Data from BS 4500

Fig. 12.9 Metric fits

12.6 Surface finish

Surface finish, more correctly termed 'surface texture', is important for all machine elements that are produced by machining processes such as turning, grinding, shaping, or honing. This applies to surfaces which are flat or cylindrical. Surface texture is covered by its own technical standard: ASME/ANSI B46.1: 1995: *Surface Texture.* It is measured using the parameter R_a which is a measurement of the average distance between the median line of the surface profile and its peaks and troughs, measured in microinches (μ in). There is another system from a comparable European standard, DIN ISO 1302, which uses a system of N-numbers –

it is simply a different way of describing the same thing.

12.6.1 Choice of surface finish: approximations

Basic surface finish designations are:

- Rough turned, with visible tool marks: 500 µin R_a (12.5 µm or N10)
- Smooth machined surface: 125 µin R_a (3.2 µm or N8)
- Static mating surfaces (or datums): 63 µin R_a (1.6 µm or N7)
- Bearing surfaces: 32 µin R_a (0.8 µm or N6)
- Fine 'lapped' surfaces: 1 µin R_a (0.025 µm or N1)

Figure 12.10 shows comparison between the different methods of measurement.

Finer finishes can be produced but are more suited for precision application such as instruments. It is good practice to specify the surface finish of close-fitting surfaces of machine elements, as well as other ASME/ANSI Y 14.5.1 parameters such as squareness and parallelism.

← Fine finish Rough finish →

R_a (µm) BS1134	0.025	0.05	0.1	0.2	0.4	0.8	1.6	3.2	6.3	12.5	25	50
R_a (µinch) ANSI B46.1	1	2	4	8	16	32	63	125	250	500	1000	2000
N-grade DIN ISO 1302	N1	N2	N3	N4	N5	N6	N7	N8	N9	N10	N11	N12

Ground finishes

Smooth turned

Medium turned

Seal-faces and running surfaces

Rough turned finish

A prescribed surface finish is shown on a drawing as 16 (surface finish symbol) – on a metric drawing this means 1.6µm R_a

Fig. 12.10 Surface measurement

12.7 Computer aided engineering

Computer Aided Engineering (CAE) is the generic name given to a collection of computer aided techniques used in aeronautical and other types of mechanical engineering.

Computer Aided Engineering (CAE) comprises:

- **CAD: Computer Aided Design (or Drafting)**
  - *Computer aided design* is the application of computers to the conceptual/design part of the engineering process. It includes analysis and simulation.
  - *Computer aided drafting* is the application of computer technology to the production of engineering drawings and images.
- **CAM: Computer Aided Manufacture** relates to the manufacture of a product using computer-controlled machine tools of some sort.
- **MRP: Materials Requirements Planning/ Manufacturing Resource Planning:** defines when a product is made, and how this fits in with the other manufacturing schedules in the factory.
- **CIM: Computer Integrated Manufacture** is the integration of all the computer-based techniques used in the design and manufacture of engineering products.

Figure 12.11 shows a general representation of how these techniques fit together.

12.7.1 CAD software

CAD software exists at several levels within an overall CAE system. It has different sources, architecture and problems. A typical structure is:

- *Level A: Operating systems*: Some are manufacturer-specific and tailored for use on their own systems.
- *Level B: Graphics software*: This governs the type and complexity of the graphics that both the CAD and CAM elements of a CAE system can display.

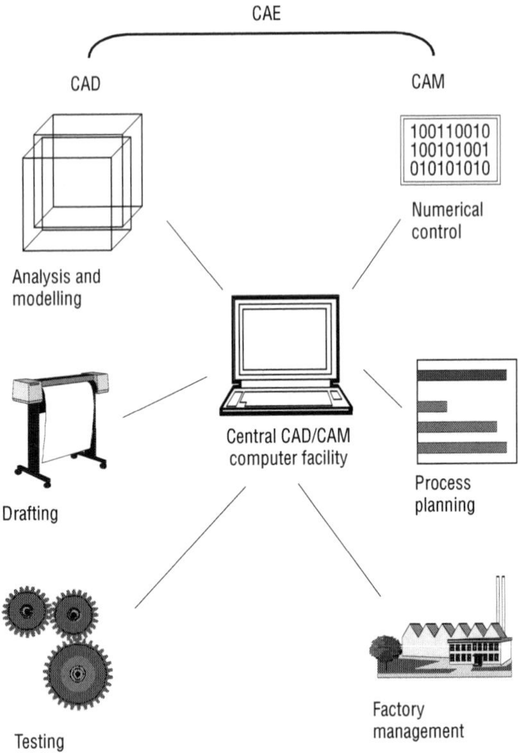

Fig. 12.11 CAE, CAD and CAM

- *Level C: Interface/Exchange software*: This comprises the common software that will be used by all the CAD/CAM application, e.g. user interface, data exchange etc.
- *Level D: Geometric modelling programs*: Most of these are designed to generate an output which can be translated into geometric form to guide a machine tool.
- *Level E: Applications software*: This is the top level of vendor-supplied software and includes drafting, and analysis/simulation facilities.
- *Level F: User-defined software*: Many systems need to be tailored before they can become truly user-specific. This category

contains all the changes required to adapt vendor software for custom use.

12.7.2 Types of modelling

CAD software packages are divided into those that portray two-dimensional or three-dimensional objects. 3D packages all contain the concept of an *underlying model.* There are three basic types as shown in Figure 12.12

Wireframe models

Although visually correct these do not contain a full description of the object. They contain no information about the surfaces and cannot differentiate between the inside and outside. They cannot be used to link to a CAM system.

Surface models

Surface models are created (conceptually) by stretching a two-dimensional 'skin' over the

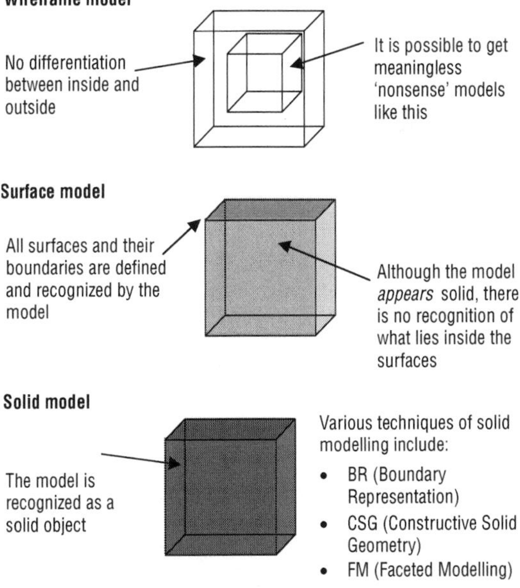
Wireframe model
No differentiation between inside and outside
It is possible to get meaningless 'nonsense' models like this
Surface model
All surfaces and their boundaries are defined and recognized by the model
Although the model appears solid, there is no recognition of what lies inside the surfaces
Solid model
The model is recognized as a solid object
Various techniques of solid modelling include:
• BR (Boundary Representation)
• CSG (Constructive Solid Geometry)
• FM (Faceted Modelling)

Fig. 12.12 Types of modelling

edges of a wireframe to define the surfaces. They can therefore define structure boundaries, but cannot distinguish a hollow object from a solid one. Surface models can be used for geometric assembly models etc., but not analyses which require the recognition of the solid properties of a body (finite element stress analysis, heat transfer etc.).

Solid models

Solid models provide a full three-dimensional geometrical definition of a solid body. They require large amounts of computer memory for definition and manipulation but can be used for finite element applications. Most solid modelling systems work by assembling a small number of 'building block' reference shapes.

12.7.3 Finite Element (FE) analysis

FE software is the most widely used type of engineering analysis package. The basic idea is that large three-dimensional areas are subdivided into small triangular or quadrilateral (planar) or hexahedral (three-dimensional) *elements* then subject a to solution of multiple simultaneous equations. The general process is loosely termed *mesh generation*. There are four types which fall into the basic category.

- **Boundary Element Modelling (BEM):** This is a simplified technique used for linear or static analyses where boundary conditions (often assumed to be at infinity) can be easily set. It is useful for analysis of cracked materials and structures.
- **Finite Element Modelling (FEM):** The technique involves a large number of broadly defined (often symmetrical) elements set between known boundary conditions. It requires large amounts of computing power.
- **Adaptive Finite Element Modelling (AFEM):** This is a refinement of FEM in which the element 'mesh' is more closely

defined in critical areas. It produces better accuracy.

- **Finite Difference Method:** A traditional method which has now been superseded by other techniques. It is still used in some specialized areas of simulation in fluid mechanics.

12.7.4 Useful references

Standards: Limits, tolerances and surface texture

1. ANSI Z17.1: 1976: *Preferred numbers.*
2. ANSI B4.2: 1999: *Preferred metric limits and fits.*
3. ANSI B4.3: 1999: *General tolerances for metric dimensioned products.*
4. ANSI/ASME Y14.5.1 M: 1999: *Dimensioning and Tolerances – mathematical definitions of principles.*
5. ASME B4.1: 1999: *Preferred limits and fits for cylindrical parts.*
6. ASME B46.1: 1995: *Surface texture (surface roughness, waviness and lay)*
7. ISO 286–1: 1988: *ISO system of limits and fits.*

Standards: Screw threads

1. ASME B1.1: 1989: *Unified inch screw threads (UN and UNR forms).*
2. ASME B1.2: 1991: *Gauges and gauging for unified screw threads.*
3. ASME B1.3M: 1992: *Screw thread gauging systems for dimensional acceptability – inch and metric screws.*
4. ASME B1.13: 1995: *Metric screw threads.*
5. ISO 5864: 1993: *ISO inch screw threads – allowances and tolerances.*

Websites

1. For a general introduction to types of CAD/CAM go to 'The Engineering Zone' at www.flinthills.com/~ramsdale/EngZone/cadcam.htm. This site also contains lists of links to popular journal sites such as *CAD/CAM magazine* and *CAE magazine.*

2. 'Finite Element Analysis World' includes listings of commercial software. Go to: www.comco.com/feaworld/feaworld.html.
3. For a general introduction to Computer Integrated Manufacture (CIM) go to: www.flinthills.com/~ramsdale/EngZone/cim.htm.
4. *The International Journal of CIM*, go to: www.tandfdc.com/jnls/cim.htm.
5. For an online introductory course on CIM, go to: www.management.mcgill.ca/course/msom/MBA/mgmt-tec/students/cim/TEST.htm.
6. For a list of PDM links, go to: www.flinthills.com/~ramsdale/EngZone/pdm.htm.
7. The PDM Information Center PDMIC is a good starting point for all PDM topics. Go to: www.pdmic.com/. For a bibliography listing, go to: www.pdmic.com/bilbliographies/index.html.

Section 13

Reference sources

13.1 Websites

Table 13.1 provides a list of useful aeronautical websites.

13.2 Fluid mechanics and aerodynamics

Flight Dynamic Principles. M.V. Cook. ISBN 0-340-63200-3. Arnold 1997.

Performance and Stability of Aircraft. J.B. Russell. ISBN 0-340-63170-8. Arnold 1996.

Aerodynamics for Engineering Students, 4th ed. E.L. Houghton, P.W. Carpenter. ISBN 0-340-54847-9. Arnold 1993.

Introduction to Fluid Mechanics. Y. Nakayama, R.F. Boucher. ISBN 0-340-67649-3. Arnold 1999.

Fluid Mechanics: An Interactive Text. J.A. Liggett, D.A. Caughey. ISBN 0-7844-0310-4. AIAA: 1998. This is a multimedia CD-ROM for fluid mechanics.

13.3 Manufacturing/materials/structures

Composite Airframe Structures, Michael C.Y. Niu, Conmilit Press Ltd, Hong Kong, 1992.

D.H. Middleton, 'The first fifty years of composite materials in aircraft construction', *Aeronautical Journal*, March 1992, pp. 96–104

Aerospace Thermal Structures and Materials for a New Era. ISBN 1-56347-182-5. AIAA publication 1995.

Aircraft Structures for Engineering Students, 3rd ed. T.H.G. Megson. ISBN 0-340-70588-4. Arnold 1999.

Table 13.1 Useful aeronautical websites

Advisory Group for Aerospace Research and Development (AGARD)	http://www.wkap.nl/natopco/pco_aga.htm
Aerospace Engineering Test Establishment (AETE)	http://www.achq.dnd.ca/aete/index.htm
Aerospace Technical Services (Australia)	http://www.aerospace.com.au/
Aerospatiale	http://www.aerospatiale.fr/
Air Force Development Test Center (AFDTC)	http://www.eglin.af.mil/afdtc/afdtc.html
Air Force Flight Test Center (AFFTC)	http://www.edwards.af.mil/
Air Force Operational Test and Evaluation Center (AFOTEC)	http//www.afotec.af.mil/
Airbus Industrie	http://www.airbus.com/
Aircraft Data	http://www.arnoldpublishers.com/aerodata/appendices/data -a/default.htm
Aircraft Locator – Manufacturer Index	http://www.brooklyn cuny.edu/rec/air/museums/manufact/manufact.html
Airports Council International (ACI)	http://www.airports.org/
Allied Signal	http://www.alliedsignal.com/
American and Canadian Aviation Directory	http://hitech.superlink/net/av/
American Institute of Aeronautics and Astronautics (AIAA)	http://www.aiaa.org/
American Society of Mechanical Engineering	http://www.asme.org/
Army Aviation Technical Test Center (ATTC)	http://www.attc.army.mil/
Arnold Engineering Development Center (AEDC)	http://info.arnold.af.mil/
Australian Centre for Test and Evaluation	http://www.acte.unisa.edu.au/weblinks.htm

BOEING Technology Services	http://www.boeing.com/bts/
British Aerospace	http://www.bae.co.uk/
CASA	http://www.casa.es/
Civil Aviation Authority (CAA)	http://www.caa.co.uk/
Daimler Chrysler Aerospace	http://www.dasa.com/
Defence Evaluation & Research Agency (DERA) United Kingdom	http://www.dera.gov.uk/
Defence Technical Information Center (DTIC)	http://www.dtic.dla.mil/
DefenseLINK	http://www.dtic.dla.mil/defenselink/index.html
Director, Test, Systems Engineering and Evaluation (DTSE&E)	http://www.acq.osd.mil/te/index.html
Directory of Technical Engineering and Science Societies and Organizations	http://www.techexpo.com/tech_soc.html
DLR – German Aerospace Research Establishment	http://www.dlr.de/
DoD-TECNET: The Test and Evaluation Community Network	http://www.tecnet0.jcte.jcs.mil:9000/index.html
Dryden Flight Research Center (DFRC) – NASA	http://www.dfrc.nasa.gov/
Edinburgh Engineering Virtual Library (EEVL)	http://www.eevl.ac.uk/
Electronic Systems Center (ESC)	http://www.hanscom.af.mil/
Engine Data	http://www.arnoldpublishers.com/aerodata/appendices/data-b/default.htm
Experimental Aircraft Association (EAA)	http://www.eaa.org/
Federal Aviation Administration	http//www.faa.gov/
National Aeronautical and Space Administration (NASA)	http://www.nasa.gov/
Flight Test Safety Committee (FTSC)	http://www.netport.com/setp/ftsc/index.html
Fokker	http://www.fokker.com/

Table 13.1 ***Continued***

General Electric Aircraft Engines	http://www.ge.com/aircraftengines/
Institution of Electrical and Electronic Engineers (IEEE)	http://www.ieee.org/
Institution of Mechanical Engineers (IMechE)	http://www.imeche.org.uk
International Federation of Airworthiness	http://www.ifairworthy.org/
International Test and Evaluation Association (ITEA)	http://www.itea.org/
International Test Pilots School (ITPS), United Kingdom	http://www.itps.uk.com/
Major Range Test Facilities Base (MRTFB)	http://www.acq.osd.mil/te/mrtfb.html
McDonnell Douglas Corporation	http//www.mdc.com/
National Aerospace Laboratory (Netherlands)	http://www.nlr.nl/
National Test Pilot School (NTPS)	http://www.ntps.com/
Naval Air Warfare Center – Aircraft Division (NAWCAD)	http://www.nawcad.navy.mil/
Naval Air Warfare Center – US Navy Flight Test	http://www.flighttest.navair.navy.mil/
Naval Air Warfare Center – Weapons Division (NAWCWPNS)	http://www.nawcwpns.namy.mil/
Nellis Air Force Base	http://www.nellis.af.mil/
North Atlantic Treaty Organization (NATO)	http://www.nato.int/
Office National d'Études et de Recherches Aérospatiales (France)	http://www.onera.fr/
Office of the Director; Operational Test & Evaluation	http://www.dote.osd.mil/
Pratt & Witney	http://www.pratt-whitney.com/
Rolls-Royce	http://www.rolls-royce.co.uk/
Royal Aeronautical Society	http://www.raes.org.uk/default.htm

Society of Automotive Engineers (SAE)	http://www.sae.org/
Society of Experimental Test Pilots (SETP)	http://www.netport.com/setp/
Society of Flight Test Engineers (SFTE), North Texas Chapter	http://www.rampages.onramp.net/~sfte/
United States Air Force Museum	http://www.wpafb.af.mil/museum/index.htm
University Consortium for Continuing Education (UCCE)	http://www.ucce.edu/
University of Tennessee Space Institute, Aviation Systems Department	http://www.utsi.edu/Academic/graduate.html
Virginia Tech Aircraft Design Information Sources	http://www.aoe.vt.edu/Mason/ACinfoTOC.html
VZLYOT Incorporated (Russia)	http://www.dsuper.net/~vzlyot/

Edinburgh (UK) Engineering Virtual Library (EEVL)
EEVL is one of the best 'gateway' sites to quality aeronautical engineering information on the internet. It contains:

The EEVL catalogue:	Descriptions and links to more than 600 aeronautical and 4500 engineering-related websites which can be browsed by engineering subject or resource type (journals, companies, institutions etc.).
Engineering newsgroups:	Access to over 100 engineering newsgroups.
Top 25 and 250 sites:	Records of the most visited engineering websites.

Access the EEVL site at http:/www.eevl.ac.uk

13.4 Aircraft sizing/multidisciplinary design

C. Bil, 'ADAS: A Design System for Aircraft Configuration Development', AIAA Paper No. 89-2131. July 1989.

S. Jayaram, A. Myklebust and P. Gelhausen, 'ACSYNT – A Standards-Based System for Parametric Computer Aided Conceptual Design of Aircraft', AIAA Paper 92-1268, Feb. 1992.

Ilan Kroo, Steve Altus, Robert Braun, Peter Gage and Ian Sobieski, 'Multidisciplinary Optimization Methods for Aircraft Preliminary Design', AIAA Paper 94-4325, 1994.

P.J. Martens, 'Airplane Sizing Using Implicit Mission Analysis', AIAA Paper 94-4406, Panama City Beach, Fl., September 1994.

Jane Dudley, Ximing Huang, Pete MacMillin, B. Grossman, R.T. Haftka and W.H. Mason, 'Multidisciplinary Optimization of the High-Speed Civil Transport', AIAA Paper 95–0124, January 1995.

The anatomy of the airplane, 2nd ed. D. Stinton. ISBN 1-56347-286-4. Blackwell, UK: 1998.

Civil jet aircraft design. L.R. Jenkinson, P. Simpkin and D. Rhodes. ISBN 0-340-74152. Arnold 1999.

13.5 Helicopter technology

Basic Helicopter Aerodynamics. J. Seddon. ISBN 0-930403-67-3. Blackwell UK: 1990.

The Foundations of Helicopter Flight. S. Newman. ISBN 0-340-58702-4. Arnold 1994.

13.6 Flying wings

The Flying Wings of Jack Northop. Gary R. Pape with Jon M. Campbell and Donna Campbell, Shiffer Military/Aviation History, Atglen, PA, 1994.

Tailless Aircraft in Theory and Practice. Karl Nickel and Michael Wohfahrt, AIAA, Washington, 1994.

David Baker, 'Northrop's big wing – the B-2' *Air International*, Part 1, Vol. 44, No. 6, June 1993, pp. 287–294.

Northrop B-2 Stealth Bomber. Bill Sweetman. Motorbooks Int'l. Osceola, WI, 1992.

13.7 Noise

Aircraft Noise. Michael J. T. Smith, Cambridge University Press, Cambridge, 1989.

E.E. Olson, 'Advanced Takeoff Procedures for High-Speed Civil Transport Community Noise Reduction', SAE Paper 921939, Oct. 1992.

13.8 Landing gear

Chai S. and Mason W.H. 'Landing Gear Integration in Aircraft Conceptual Design,' AIAA Paper 96–4038, Proceedings of the 6th AIAA/NASA/ISSMO Symposium on Multidisciplinary Analysis and Optimization, Sept. 1996. pp. 525–540. Acrobat format.

S.J. Greenbank, 'Landing Gear – The Aircraft Requirement', *Proceedings of Institution of Mechanical Engineers* (UK), Vol. 205, 1991, pp.27–34.

Airframe Structural Design. M.C.Y. Niu. Conmilit Press, Ltd, Hong Kong, 1988. This book contains a good chapter on landing gear design.

S.F.N. Jenkins. 'Landing Gear Design and Development', Institution of Mechanical Engineers (UK), proceedings, part G1, *Journal of Aerospace Engineering*, Vol. 203, 1989.

13.9 Aircraft operations

Aircraft Data for Pavement Design. American Concrete Pavement Association, 1993.

Airport Engineering, 3rd ed. Norman Ashford and Paul H. Wright. John Wiley & Sons, Inc., 1992.

13.10 Propulsion

Walter C. Swan and Armand Sigalla, 'The Problem of Insalling a Modern High Bypass Engine on a Twin Jet Transport Aircraft', in *Aerodynamic Drag*, AGARD CP-124, April 1973.

The Development of Piston Aero Engines. Bill Gunston. Patrick Stephens Limited, UK, 1993.

Aircraft Engine Design. J.D. Maltingly, W.H. Heiser, D.H. Daley. ISBN 0-930403-23-1. AIAA Education Series, 1987.

Appendix 1: Aerodynamic stability and control derivatives

Table A1.1 Longitudinal aerodynamic stability derivatives

Dimensionless	*Multiplier*	*Dimensional*
X_M	$\frac{1}{2}\rho V_0 S$	$\mathring{X}_u$
X_w	$\frac{1}{2}\rho V_0 S$	$\mathring{X}_w$
$X_{\dot{w}}$	$\frac{1}{2}\rho S\bar{\bar{c}}$	$\mathring{X}_{\dot{w}}$
X_q	$\frac{1}{2}\rho V_0 S\bar{\bar{c}}$	$\mathring{X}_q$
Z_M	$\frac{1}{2}\rho V_0 S$	$\mathring{Z}_u$
Z_w	$\frac{1}{2}\rho V_0 S$	$\mathring{Z}_w$
$Z_{\dot{w}}$	$\frac{1}{2}\rho S\bar{\bar{c}}$	$\mathring{Z}_{\dot{w}}$
Z_q	$\frac{1}{2}\rho V_0 S\bar{\bar{c}}$	$\mathring{Z}_q$
M_u	$\frac{1}{2}\rho V_0 S\bar{\bar{c}}$	$\mathring{M}_u$
M_w	$\frac{1}{2}\rho V_0 S\bar{\bar{c}}$	$\mathring{M}_w$
$M_{\dot{w}}$	$\frac{1}{2}\rho S\bar{\bar{c}}^2$	$\mathring{M}_{\dot{w}}$
M_q	$\frac{1}{2}\rho V_0 S\bar{\bar{c}}^2$	$\mathring{M}_q$

Table A1.2 Longitudinal control derivatives

Dimensionless	*Multiplier*	*Dimensional*
X_η	$\frac{1}{2}\rho V_0^2 S$	$\mathring{X}_\eta$
Z_η	$\frac{1}{2}\rho V_0^2 S$	$\mathring{Z}_\eta$
M_η	$\frac{1}{2}\rho V_0^2 S\bar{\bar{c}}$	$\mathring{M}_\eta$
X_τ	1	$\mathring{X}_\tau$
Z_τ	1	$\mathring{Z}_\tau$
M_τ	$\bar{\bar{c}}_\tau$	$\mathring{M}_\tau$

Table A1.3 Lateral aerodynamic stability derivatives

Dimensionless	*Multiplier*	*Dimensional*
Y_v	$\frac{1}{2}\rho V_0 S$	$\mathring{Y}_v$
Y_p	$\frac{1}{2}\rho V_0 Sb$	$\mathring{Y}_p$
Y_r	$\frac{1}{2}\rho V_0 Sb$	$\mathring{Y}_r$
L_v	$\frac{1}{2}\rho V_0 Sb$	$\mathring{L}_v$
L_p	$\frac{1}{2}\rho V_0 Sb^2$	$\mathring{L}_p$
L_r	$\frac{1}{2}\rho V_0 Sb^2$	$\mathring{L}_r$
N_v	$\frac{1}{2}\rho V_0 Sb$	$\mathring{N}_v$
N_p	$\frac{1}{2}\rho V_0 Sb^2$	$\mathring{N}_p$
N_r	$\frac{1}{2}\rho V_0 Sb^2$	$\mathring{N}_r$

Table A.14 Lateral aerodynamic control derivatives

Dimensionless	*Multiplier*	*Dimensional*
Y_ξ	$\frac{1}{2}\rho V_0^2 S$	$\mathring{Y}_\xi$
L_ξ	$\frac{1}{2}\rho V_0^2 Sb$	$\mathring{L}_\xi$
N_ξ	$\frac{1}{2}\rho V_0^2 Sb$	$\mathring{N}_\xi$
Y_ζ	$\frac{1}{2}\rho V_0^2 S$	$\mathring{Y}_\zeta$
L_ζ	$\frac{1}{2}\rho V_0^2 Sb$	$\mathring{L}_\zeta$
N_ζ	$\frac{1}{2}\rho V_0^2 Sb$	$\mathring{N}_\zeta$

Appendix 2: Aircraft response transfer functions

Table A2.1 Longitudinal response transfer functions

η is elevator input.

Common denominator polynomial $\Delta(s) = as^4 + bs^3 + cs^2 + ds + e$

a $mI_y (m - \mathring{Z}_{\dot{w}})$

b $I_y (\mathring{X}_u \mathring{Z}_{\dot{w}} - \mathring{X}_{\dot{w}} \mathring{Z}_u) - mI_Y (\mathring{X}_u + \mathring{Z}_w) - mM_{\dot{w}} (\mathring{Z}_q + mU_e) - m\mathring{M}_q (m - \mathring{Z}_{\dot{w}})$

c $I_y (\mathring{X}_u \mathring{Z}_{\dot{w}} - \mathring{X}_w \mathring{Z}_u) + (\mathring{X}_u \mathring{M}_{\dot{w}} - \mathring{X}_w \mathring{M}_u)(\mathring{Z}_q + mU_e)$
$+ \mathring{Z}_u (\mathring{X}_{\dot{w}} \mathring{M}_q - \mathring{X}_q \mathring{M}_{\dot{w}}) + (\mathring{X}_u \mathring{M}_q - \mathring{X}_q \mathring{M}_u)(m - \mathring{Z}_{\dot{w}})$
$+ m(\mathring{M}_q \mathring{Z}_w - \mathring{M}_w \mathring{Z}_q) + mW_e (\mathring{M}_{\dot{w}} \mathring{Z}_u - \mathring{M}_u \mathring{Z}_{\dot{w}})$
$+ m^2(\mathring{M}_{\dot{w}}\, g \sin \theta_e - u_e \mathring{M}_w)$

d $(\mathring{X}_u \mathring{M}_w - \mathring{X}_w \mathring{M}_u)(\mathring{Z}_q + mU_e)$
$+ (\mathring{M}_u \mathring{Z}_w - \mathring{M}_w \mathring{Z}_u)(\mathring{X}_q\, mW_e) + \mathring{M}_q (\mathring{X}_w \mathring{Z}_u - \mathring{X}_u \mathring{Z}_w)$
$+ mg \cos\theta_e (\mathring{M}_{\dot{w}} \mathring{Z}_u + \mathring{M}_u (m - \mathring{Z}_{\dot{w}})) + mg \sin \theta_e (\mathring{X}_{\dot{w}} \mathring{M}_u - \mathring{X}_u \mathring{M}_w + m\mathring{M}_w)$
$+ mg \sin \theta_e (\mathring{X}_w \mathring{M}_u - \mathring{X}_u \mathring{M}_w) + mg \cos\theta_e (\mathring{M}_w \mathring{Z}_u - \mathring{M}_u \mathring{Z}_w)$

e $mg \sin \theta_e (\mathring{X}_w \mathring{M}_u - \mathring{X}_u \mathring{M}_w) + mg \cos\theta_e (\mathring{M}_w \mathring{Z}_u - \mathring{M}_u \mathring{Z}_w)$

Numerator polynomial $N_3^{\mu}(s) = as^2 + bs^2 + cs + d$

a $I_y (\mathring{X}_{\dot{w}} \mathring{Z}_\eta + \mathring{X}_\eta (m - \mathring{Z}_{\dot{w}}))$

b $\mathring{X}_\eta (-I_y \mathring{Z}_w + mU_e) - \mathring{M}_q (m - \mathring{Z}_{\dot{w}}))$
$+ \mathring{Z}_\eta (I_y \mathring{X}_w - \mathring{X}_{\dot{w}} \mathring{M}_q + \mathring{M}_{\dot{w}} (\mathring{X}_q - mW_e))$
$+ \mathring{M}_\eta ((\mathring{X}_q - mW_e)(m - \mathring{Z}_{\dot{w}}) + \mathring{X}_{\dot{w}} (\mathring{Z}_q + mU_e))$

c $\mathring{X}_\eta (\mathring{Z}_w \mathring{M}_q - (\mathring{M}_w (\mathring{Z}_q + mU_e) + mg \sin \theta_e \mathring{M}_{\dot{w}})$
$+ \mathring{Z}_\eta (\mathring{M}_w (\mathring{X}_q - mW_e) - \mathring{X}_w \mathring{M}_q - mg \cos\theta_e \mathring{M}_{\dot{w}})$
$+ \mathring{M}_\eta (\mathring{X}_w (\mathring{Z}_q + mU_e) - \mathring{Z}_w (\mathring{X}_q - mW_e) - mg \cos\theta_e (m - \mathring{Z}_{\dot{w}}) - mg \sin \theta_e \mathring{X}_{\dot{w}})$

d $\mathring{X}_\eta \mathring{M}_w\, mg \sin \theta_e - \mathring{Z}_\eta \mathring{M}_w\, mg \cos\theta_e + \mathring{M}_\eta (\mathring{Z}_w\, mg \cos \theta_e - \mathring{X}_w\, mg \sin \theta_e)$

Table A2.2 Lateral-directional response transfer functions in terms of dimensional derivatives

ξ is aileron input

Demoninator polynomial $\Delta(s) = s(as^4 + bs^3 + cs^2 + ds + e)$

a $m(I_xI_z - I^2_{xz})$

b $-\mathring{Y}_v (I_xI_z - I^2_{xz}) - m(I_x \mathring{N}_r + I_{xz} \mathring{L}_r) - m(I_z \mathring{L}_p + I_{xz} \mathring{N}_p)$

c $\mathring{Y}_v (I_x \mathring{N}_r + I_{xz} \mathring{L}_r) + \mathring{Y}_v (I_z \mathring{L}_p + I_{xz} \mathring{N}_p) - (\mathring{Y}_p + mW_e)(I_z \mathring{L}_v + I_{xz} \mathring{N}_v)$
$- (\mathring{Y}_r - mU_e)(I_x \mathring{N}_v + I_{xz} \mathring{L}_v) + m(\mathring{L}_p \mathring{N}_r - \mathring{L}_r \mathring{N}_p)$

d $- (\mathring{Y}_v (\mathring{L}_r \mathring{N}_p - \mathring{L}_p \mathring{N}_r) + (\mathring{Y}_p + mW_e)(\mathring{L}_v \mathring{N}_r - \mathring{L}_r \mathring{N}_v)$
$(\mathring{Y}_r - mU_e)(\mathring{L}_p \mathring{N}_v - \mathring{L}_v \mathring{N}_p)$
$- mg \cos\theta_e (I_z \mathring{L}_v + I_{xz} \mathring{N}_v) - mg \sin\theta_e (I_x \mathring{N}_v + I_{xz} \mathring{L}_v)$

e $mg \cos\theta_e (\mathring{L}_v \mathring{N}_r - \mathring{L}_r \mathring{N}_v) + mg \sin\theta_e (\mathring{L}_p \mathring{N}_v - \mathring{L}_v \mathring{N}_p)$

Numerator polynomial $N^v_\xi (s) = s(as^3 + bs^2 + cs + d)$

a $\mathring{Y}_\xi (I_xI_z - I^2_{xz})$

b $\mathring{Y}_\xi (-I_x \mathring{N}_r - I_z \mathring{L}_p - I_{xz} (\mathring{L}_r \mathring{N}_p)) + \mathring{L}_\xi (I_z(\mathring{Y}_p + mW_e) + I_{xz} (\mathring{Y}_r - mU_e))$
$+ \mathring{N}_\xi (I_x(\mathring{Y}_r - mU_e) + I_{xz} (\mathring{Y}_p + mW_e))$

c $\mathring{Y}_\xi (\mathring{L}_p \mathring{N}_r - \mathring{L}_r \mathring{N}_p)$
$+ \mathring{L}_\xi (\mathring{N}_p (\mathring{Y}_r - mU_e) - \mathring{N}_r (\mathring{Y}_p + mW_e) + mg(I_z \cos\theta_e + I_{xz} \sin\theta_e))$
$+ \mathring{N}_\xi (\mathring{L}_r (\mathring{Y}_p - mW_e) - \mathring{L}_p (\mathring{Y}_r + mU_e) + mg(I_x \sin\theta_e + I_{xz} \cos\theta_e))$

d $\mathring{L}_\xi (\mathring{N}_p mg \sin\theta_e - \mathring{N}_r mg \cos\theta_e) + \mathring{N}\xi (\mathring{L}_r mg \cos\theta_e - \mathring{L}_p mg \cos\theta_e)$

Appendix 3: Approximate expressions for the dimensionless aerodynamic stability and control derivatives

Table A3.1 Longitudinal aerodynamic stability derivatives

Small perturbation derivatives referred to aircraft wind axes

Derivative	*Description*	*Expression*	*Comments*
X_u	Axial force due to velocity	$-2C_D - V_0 \dfrac{\partial C_D}{\partial V} + \dfrac{1}{\frac{1}{2}\rho V_0 S} \dfrac{\partial \tau}{\partial V}$	Drag and thrust effects due to velocity perturbation
X_w	Axial force due to incidence	$C_L - \dfrac{\partial C_D}{\partial \alpha}$	Lift and drag effects due to incidence perturbation
X_q	Axial force due to pitch rate	$-\overline{V}_r \dfrac{\partial C_{D_T}}{\partial \alpha_T}$	Tailplane drag effect, usually negligible
$X_{\dot{w}}$	Axial force due to downwash lag	$-\overline{V}_r \dfrac{\partial C_{D_T}}{\partial \alpha_T} \dfrac{d\epsilon}{d\alpha} \equiv X_q \dfrac{d\epsilon}{d\alpha}$	Tailplane drag due to downwash lag effect (added mass effect)
Z_u	Normal force due to velocity	$-2C_L - V_0 \dfrac{\partial C_L}{\partial V}$	Lift effects due to velocity perturbation
Z_w	Normal force due to 'incidence'	$-C_D - \dfrac{\partial C_L}{\partial \alpha}$	Lift and drag effects due to incidence perturbation

Z_q	Normal force due to pitch rate	$-\overline{V}_r\alpha_1$	Tailplane lift effect
$Z_{\mathring{w}}$	Normal force due to downwash lag	$-\overline{V}_r\alpha_1 \frac{d\epsilon}{d\alpha} = Z_q \frac{d\epsilon}{d\alpha}$	Tailplane lift due to downwash lag effect (added mass effect)
M_u	Pitching moment due to velocity	$V_0 \frac{\partial C_m}{\partial V}$	Mach dependent, small at low speed
M_w	Pitching moment due to 'incidence'	$\frac{dC_m}{d\alpha} = -\alpha K_n$	Pitch stiffness, dependent on static margin
M_q	Pitching moment due to pitch rate	$-\overline{V}_T \frac{\mathrm{l_T}}{\overline{\overline{c}}} \equiv Z_q \frac{l_T}{\overline{\overline{c}}}$	Pitch damping, due mainly to tailplane
$M_{\mathring{w}}$	Pitching moment due to downwash lag	$-\overline{V}_T\alpha_1 \frac{l_T}{\overline{\overline{c}}} \frac{d}{d\alpha} \equiv M_q \frac{d\varepsilon}{d\alpha}$	Pitch damping, due to downwash lag effect at tailplane

Table A3.2 Small perturbation derivatives referred to aircraft wind axes

Derivative	*Description*	*Expression*		*Comments*
Y_v	Sideforce due to sideslip		$\left(\frac{S_B}{S} y_B - \frac{S_F}{S}\alpha_{1_F}\right)$	Always negative and hence stabilizing
L_v	Rolling moment due to sideslip	(i) wing with dihedral	$-\frac{1}{Ss}\int_0^s c_y a_y \Gamma y dy$	Lateral static stability, determined by total dihedral effect. Most accessible approximate contribution is given
		(ii) wing with aft sweep	$-\frac{2C_L \tan \Lambda_{1/4}}{Ss}\int_0^s c_y y dy$	
		(iii) fin contribution	$a_{1_F}\overline{V}_F \frac{h_F}{l_F}$	
N_v	Yawing moment due to sideslip	(i) fin contribution	$a_{1_F}\overline{V}_F$	Natural weathercock stability, dominated by fin effect
Y_p	Sideforce due to roll rate	(i) fin contribution	$-\frac{1}{Sb}\int_0^{H_F} a_h c_h h dh$	Fin effect dominates, often negligible

L_p	Rolling moment due to roll rate	(i) wing contribution	$-\frac{1}{2Ss^2}\int_0^s (a_y + C_{D_y})c_y y^2 dy$	Roll damping wing effects dominate but fin and tailplane contribute
N_p	Yawing moment due to roll rate	(i) wing contribution	$-\frac{1}{2Ss^2}\int_0^s \left(C_{L_y} - \frac{dC_D}{da_y}\right)c_y y^2 dy$	
Y_r	Sideforce due to yaw rate	*(i) fin contribution*	$\overline{V}_F a_{1_F}$	Many contributions, but often negligible
L_r	Rolling moment due to yaw rate	(i) wing contribution	$\frac{1}{Ss^2}\int_0^s C_{L_y} c_y y^2 dy$	
		(ii) fin contribution	$a_{1_F}\overline{V}_F \frac{h_F}{b} \equiv -L_{v(\text{fin})}\frac{l_F}{b}$	
N_r	Yawing moment due to yaw rate	(i) wing contribution	$\frac{1}{Ss^2}\int_0^s C_{D_y} c_y y^2 dy$	Yaw damping, for large aspect ratio rectangular wing, wing contribution is approximately $C^D/_6$
		(ii) fin contribution	$a_{1_F}\overline{V}_F \frac{l_F}{b} \equiv -\frac{l_F}{b} N_{v(\text{fin})}$	

Table A3.3 Longitudinal aerodynamic control derivatives

Small perturbation derivatives referred to aircraft wind axes

Derivative	*Description*	*Expression*	*Comments*
X_η	Axial force due to elevator	$-2\dfrac{S_T}{S}k_T C_{L_T} a_2$	Usually insignificantly small
Z_η	Normal force due to elevator	$-\dfrac{S_T}{S}a_2$	
M_η	Pitching moment due to elevator	$-\overline{V}_T a_2$	Principal measure of pitch control power

Appendix 4: Compressible flow tables

Table A4.1 Subsonic flow (isentropic flow, $\gamma = 7/5$)

Notation:
M = Local flow Mach number
P/P_o = Ratio of static pressure to total pressure
ρ/ρ_o = Ratio of local flow density to stagnation density
T/T_o = Ratio of static temperature to total temperature
$\beta = (1 - M_2)$ = Compressibility factor
V/a^* = Local velocity/speed of sound at sonic point
q/P_o = Dynamic pressure/total pressure
A/A^* = Local flow area/flow area at sonic point

M	P/P_o	ρ/ρ_o	T/T_o	β	q/P_o	A/A^*	V/a^*
0.00	1.0000	1.0000	1.0000	1.0000	0.0000	–	0.0000
0.01	0.9999	1.0000	1.0000	0.9999	7.000e–5	57.8738	0.0110
0.02	0.9997	0.9998	0.9999	0.9998	2.799e–4	28.9421	0.0219
0.03	0.9994	0.9996	0.9998	0.9995	6.296e–4	19.3005	0.0329
0.04	0.9989	0.9992	0.9997	0.9992	1.119e–3	14.4815	0.0438
0.05	0.9983	0.9988	0.9995	0.9987	1.747e–3	11.5914	0.0548
0.06	0.9975	0.9982	0.9993	0.9982	2.514e–3	9.6659	0.0657
0.07	0.9966	0.9976	0.9990	0.9975	3.418e–3	8.2915	0.0766
0.08	0.9955	0.9968	0.9987	0.9968	4.460e–3	7.2616	0.0876
0.09	0.9944	0.9960	0.9984	0.9959	5.638e–3	6.4613	0.0985
0.10	0.9930	0.9950	0.9980	0.9950	6.951e–3	5.8218	0.1094
0.11	0.9916	0.9940	0.9976	0.9939	8.399e–3	5.2992	0.1204
0.12	0.9900	0.9928	0.9971	0.9928	9.979e–3	4.8643	0.1313
0.13	0.9883	0.9916	0.9966	0.9915	1.169e–2	4.4969	0.1422
0.14	0.9864	0.9903	0.9961	0.9902	1.353e–2	4.1824	0.1531
0.15	0.9844	0.9888	0.9955	0.9887	1.550e–2	3.9103	0.1639
0.16	0.9823	0.9873	0.9949	0.9871	1.760e–2	3.6727	0.1748
0.17	0.9800	0.9857	0.9943	0.9854	1.983e–2	3.4635	0.1857
0.18	0.9776	0.9840	0.9936	0.9837	2.217e–2	3.2779	0.1965
0.19	0.9751	0.9822	0.9928	0.9818	2.464e–2	3.1123	0.2074
0.20	0.9725	0.9803	0.9921	0.9798	2.723e–2	2.9635	0.2182
0.21	0.9697	0.9783	0.9913	0.9777	2.994e–2	2.8293	0.2290
0.22	0.9668	0.9762	0.9904	0.9755	3.276e–2	2.7076	0.2398
0.23	0.9638	0.9740	0.9895	0.9732	3.569e–2	2.5968	0.2506
0.24	0.9607	0.9718	0.9886	0.9708	3.874e–2	2.4956	0.2614
0.25	0.9575	0.9694	0.9877	0.9682	4.189e–2	2.4027	0.2722
0.26	0.9541	0.9670	0.9867	0.9656	4.515e–2	2.3173	0.2829
0.27	0.9506	0.9645	0.9856	0.9629	4.851e–2	2.2385	0.2936
0.28	0.9470	0.9619	0.9846	0.9600	5.197e–2	2.1656	0.3043
0.29	0.9433	0.9592	0.9835	0.9570	5.553e–2	2.0979	0.3150
0.30	0.9395	0.9564	0.9823	0.9539	5.919e–2	2.0351	0.3257

Table A4.1 ***Continued***

M	P/P_o	ρ/ρ_o	T/T_o	β	q/P_o	A/A^*	V/a^*
0.31	0.9355	0.9535	0.9811	0.9507	6.293e–2	1.9765	0.3364
0.32	0.9315	0.9506	0.9799	0.9474	6.677e–2	1.9219	0.3470
0.33	0.9274	0.9476	0.9787	0.9440	7.069e–2	1.8707	0.3576
0.34	0.9231	0.9445	0.9774	0.9404	7.470e–2	1.8229	0.3682
0.35	0.9188	0.9413	0.9761	0.9367	7.878e–2	1.7780	0.3788
0.36	0.9143	0.9380	0.9747	0.9330	8.295e–2	1.7358	0.3893
0.37	0.9098	0.9347	0.9733	0.9290	8.719e–2	1.6961	0.3999
0.38	0.9052	0.9313	0.9719	0.9250	9.149e–2	1.6587	0.4104
0.39	0.9004	0.9278	0.9705	0.9208	9.587e–2	1.6234	0.4209
0.40	0.8956	0.9243	0.9690	0.9165	0.1003	1.5901	0.4313
0.41	0.8907	0.9207	0.9675	0.9121	0.1048	1.5587	0.4418
0.42	0.8857	0.9170	0.9659	0.9075	0.1094	1.5289	0.4522
0.43	0.8807	0.9132	0.9643	0.9028	0.1140	1.5007	0.4626
0.44	0.8755	0.9094	0.9627	0.8980	0.1186	1.4740	0.4729
0.45	0.8703	0.9055	0.9611	0.8930	0.1234	1.4487	0.4833
0.46	0.8650	0.9016	0.9594	0.8879	0.1281	1.4246	0.4936
0.47	0.8596	0.8976	0.9577	0.8827	0.1329	1.4018	0.5038
0.48	0.8541	0.8935	0.9559	0.8773	0.1378	1.3801	0.5141
0.49	0.8486	0.8894	0.9542	0.8717	0.1426	1.3595	0.5243
0.50	0.8430	0.8852	0.9524	0.8660	0.1475	1.3398	0.5345
0.51	0.8374	0.8809	0.9506	0.8602	0.1525	1.3212	0.5447
0.52	0.8317	0.8766	0.9487	0.8542	0.1574	1.3034	0.5548
0.53	0.8259	0.8723	0.9468	0.8480	0.1624	1.2865	0.5649
0.54	0.8201	0.8679	0.9449	0.8417	0.1674	1.2703	0.5750
0.55	0.8142	0.8634	0.9430	0.8352	0.1724	1.2549	0.5851
0.56	0.8082	0.8589	0.9410	0.8285	0.1774	1.2403	0.5951
0.57	0.8022	0.8544	0.9390	0.8216	0.1825	1.2263	0.6051
0.58	0.7962	0.8498	0.9370	0.8146	0.1875	1.2130	0.6150
0.59	0.7901	0.8451	0.9349	0.8074	0.1925	1.2003	0.6249
0.60	0.7840	0.8405	0.9328	0.8000	0.1976	1.1882	0.6348
0.61	0.7778	0.8357	0.9307	0.7924	0.2026	1.1767	0.6447
0.62	0.7716	0.8310	0.9286	0.7846	0.2076	1.1656	0.6545
0.63	0.7654	0.8262	0.9265	0.7766	0.2127	1.1552	0.6643
0.64	0.7591	0.8213	0.9243	0.7684	0.2177	1.1451	0.6740
0.65	0.7528	0.8164	0.9221	0.7599	0.2226	1.1356	0.6837
0.66	0.7465	0.8115	0.9199	0.7513	0.2276	1.1265	0.6934
0.67	0.7401	0.8066	0.9176	0.7424	0.2326	1.1179	0.7031
0.68	0.7338	0.8016	0.9153	0.7332	0.2375	1.1097	0.7127
0.69	0.7274	0.7966	0.9131	0.7238	0.2424	1.1018	0.7223
0.70	0.7209	0.7916	0.9107	0.7141	0.2473	1.0944	0.7318
0.71	0.7145	0.7865	0.9084	0.7042	0.2521	1.0873	0.7413
0.72	0.7080	0.7814	0.9061	0.6940	0.2569	1.0806	0.7508
0.73	0.7016	0.7763	0.9037	0.6834	0.2617	1.0742	0.7602
0.74	0.6951	0.7712	0.9013	0.6726	0.2664	1.0681	0.7696
0.75	0.6886	0.7660	0.8989	0.6614	0.2711	1.0624	0.7789
0.76	0.6821	0.7609	0.8964	0.6499	0.2758	1.0570	0.7883
0.77	0.6756	0.7557	0.8940	0.6380	0.2804	1.0519	0.7975
0.78	0.6691	0.7505	0.8915	0.6258	0.2849	1.0471	0.8068
0.79	0.6625	0.7452	0.8890	0.6131	0.2894	1.0425	0.8160
0.80	0.6560	0.7400	0.8865	0.6000	0.2939	1.0382	0.8251
0.81	0.6495	0.7347	0.8840	0.5864	0.2983	1.0342	0.8343
0.82	0.6430	0.7295	0.8815	0.5724	0.3026	1.0305	0.8433
0.83	0.6365	0.7242	0.8789	0.5578	0.3069	1.0270	0.8524
0.84	0.6300	0.7189	0.8763	0.5426	0.3112	1.0237	0.8614
0.85	0.6235	0.7136	0.8737	0.5268	0.3153	1.0207	0.8704
0.86	0.6170	0.7083	0.8711	0.5103	0.3195	1.0179	0.8793

Table A4.1 *Continued*

M	P/P_o	ρ/ρ_o	T/T_o	β	q/P_o	A/A^*	V/a^*
0.87	0.6106	0.7030	0.8685	0.4931	0.3235	1.0153	0.8882
0.88	0.6041	0.6977	0.8659	0.4750	0.3275	1.0129	0.8970
0.89	0.5977	0.6924	0.8632	0.4560	0.3314	1.0108	0.9058
0.90	0.5913	0.6870	0.8606	0.4359	0.3352	1.0089	0.9146
0.91	0.5849	0.6817	0.8579	0.4146	0.3390	1.0071	0.9233
0.92	0.5785	0.6764	0.8552	0.3919	0.3427	1.0056	0.9320
0.93	0.5721	0.6711	0.8525	0.3676	0.3464	1.0043	0.9407
0.94	0.5658	0.6658	0.8498	0.3412	0.3499	1.0031	0.9493
0.95	0.5595	0.6604	0.8471	0.3122	0.3534	1.0021	0.9578
0.96	0.5532	0.6551	0.8444	0.2800	0.3569	1.0014	0.9663
0.97	0.5469	0.6498	0.8416	0.2431	0.3602	1.0008	0.9748
0.98	0.5407	0.6445	0.8389	0.1990	0.3635	1.0003	0.9833
0.99	0.5345	0.6392	0.8361	0.1411	0.3667	1.0001	0.9916

Table A4.2 Supersonic flow (isentropic flow, γ = 7/5)

Notation:
M = Local flow Mach number
P/P_o = Ratio of static pressure to total pressure
ρ/ρ_o = Ratio of local flow density to stagnation density (r/ro)
T/T_o = Ratio of static temperature to total temperature
$\beta = \sqrt{1 - M_2}$ = Compressibility factor
V/a^* = Local velocity/speed of sound at sonic point
q/P_o = Dynamic pressure/total pressure
A/A^* = Local flow area/flow area at sonic point

M	P/P_o	ρ/ρ_o	T/T_o	β	q/P_o	A/A^*	V/a^*
1.00	0.5283	0.6339	0.8333	0.0000	0.3698	1.0000	1.0000
1.01	0.5221	0.6287	0.8306	0.1418	0.3728	1.0001	1.0083
1.02	0.5160	0.6234	0.8278	0.2010	0.3758	1.0003	1.0166
1.03	0.5099	0.6181	0.8250	0.2468	0.3787	1.000	1.0248
1.04	0.5039	0.6129	0.8222	0.2857	0.3815	1.0013	1.0330
1.05	0.4979	0.6077	0.8193	0.3202	0.3842	1.0020	1.0411
1.06	0.4919	0.6024	0.8165	0.3516	0.3869	1.0029	1.0492
1.07	0.4860	0.5972	0.8137	0.3807	0.3895	1.0039	1.0573
1.08	0.4800	0.5920	0.8108	0.4079	0.3919	1.0051	1.0653
1.09	0.4742	0.5869	0.8080	0.4337	0.3944	1.0064	1.0733
1.10	0.4684	0.5817	0.8052	0.4583	0.3967	1.0079	1.0812
1.11	0.4626	0.5766	0.8023	0.4818	0.3990	1.0095	1.0891
1.12	0.4568	0.5714	0.7994	0.5044	0.4011	1.0113	1.0970
1.13	0.4511	0.5663	0.7966	0.5262	0.4032	1.0132	1.1048
1.14	0.4455	0.5612	0.7937	0.5474	0.4052	1.0153	1.1126
1.15	0.4398	0.5562	0.7908	0.5679	0.4072	1.0175	1.1203
1.16	0.4343	0.5511	0.7879	0.5879	0.4090	1.0198	1.1280
1.17	0.4287	0.5461	0.7851	0.6074	0.4108	1.0222	1.1356
1.18	0.4232	0.5411	0.7822	0.6264	0.4125	1.0248	1.1432
1.19	0.4178	0.5361	0.7793	0.6451	0.4141	1.0276	1.1508
1.20	0.4124	0.5311	0.7764	0.6633	0.4157	1.0304	1.1583
1.21	0.4070	0.5262	0.7735	0.6812	0.4171	1.0334	1.1658
1.22	0.4017	0.5213	0.7706	0.6989	0.4185	1.0366	1.1732

Table A4.2 ***Continued***

M	P/P_o	ρ/ρ_o	T/T_o	β	q/P_o	A/A^*	V/a^*
1.23	0.3964	0.5164	0.7677	0.7162	0.4198	1.0398	1.1806
1.24	0.3912	0.5115	0.7648	0.7332	0.4211	1.0432	1.1879
1.25	0.3861	0.5067	0.7619	0.7500	0.4223	1.0468	1.1952
1.26	0.3809	0.5019	0.7590	0.7666	0.4233	1.0504	1.2025
1.27	0.3759	0.4971	0.7561	0.7829	0.4244	1.0542	1.2097
1.28	0.3708	0.4923	0.7532	0.7990	0.4253	1.0581	1.2169
1.29	0.3658	0.4876	0.7503	0.8149	0.4262	1.0621	1.2240
1.30	0.3609	0.4829	0.7474	0.8307	0.4270	1.0663	1.2311
1.31	0.3560	0.4782	0.7445	0.8462	0.4277	1.0706	1.2382
1.32	0.3512	0.4736	0.7416	0.8616	0.4283	1.0750	1.2452
1.33	0.3464	0.4690	0.7387	0.8769	0.4289	1.0796	1.2522
1.34	0.3417	0.4644	0.7358	0.8920	0.4294	1.0842	1.2591
1.35	0.3370	0.4598	0.7329	0.9069	0.4299	1.0890	1.2660
1.36	0.3323	0.4553	0.7300	0.9217	0.4303	1.0940	1.2729
1.37	0.3277	0.4508	0.7271	0.9364	0.4306	1.0990	1.2797
1.38	0.3232	0.4463	0.7242	0.9510	0.4308	1.1042	1.2864
1.39	0.3187	0.4418	0.7213	0.9655	0.4310	1.1095	1.2932
1.40	0.3142	0.4374	0.7184	0.9798	0.4311	1.1149	1.2999
1.41	0.3098	0.4330	0.7155	0.9940	0.4312	1.1205	1.3065
1.42	0.3055	0.4287	0.7126	1.0082	0.4312	1.1262	1.3131
1.43	0.3012	0.4244	0.7097	1.0222	0.4311	1.1320	1.3197
1.44	0.2969	0.4201	0.7069	1.0361	0.4310	1.1379	1.3262
1.45	0.2927	0.4158	0.7040	1.0500	0.4308	1.1440	1.3327
1.46	0.2886	0.4116	0.7011	1.0638	0.4306	1.1501	1.3392
1.47	0.2845	0.4074	0.6982	1.0775	0.4303	1.1565	1.3456
1.48	0.2804	0.4032	0.6954	1.0911	0.4299	1.1629	1.3520
1.49	0.2764	0.3991	0.6925	1.1046	0.4295	1.1695	1.3583
1.50	0.2724	0.3950	0.6897	1.1180	0.4290	1.1762	1.3646
1.51	0.2685	0.3909	0.6868	1.1314	0.4285	1.1830	1.3708
1.52	0.2646	0.3869	0.6840	1.1447	0.4279	1.1899	1.3770
1.53	0.2608	0.3829	0.6811	1.1580	0.4273	1.1970	1.3832
1.54	0.2570	0.3789	0.6783	1.1712	0.4266	1.2042	1.3894
1.55	0.2533	0.3750	0.6754	1.1843	0.4259	1.2116	1.3955
1.56	0.2496	0.3710	0.6726	1.1973	0.4252	1.2190	1.4015
1.57	0.2459	0.3672	0.6698	1.2103	0.4243	1.2266	1.4075
1.58	0.2423	0.3633	0.6670	1.2233	0.4235	1.2344	1.4135
1.59	0.2388	0.3595	0.6642	1.2362	0.4226	1.2422	1.4195
1.60	0.2353	0.3557	0.6614	1.2490	0.4216	1.2502	1.4254
1.61	0.2318	0.3520	0.6586	1.2618	0.4206	1.2584	1.4313
1.62	0.2284	0.3483	0.6558	1.2745	0.4196	1.2666	1.4371
1.63	0.2250	0.3446	0.6530	1.2872	0.4185	1.2750	1.4429
1.64	0.2217	0.3409	0.6502	1.2998	0.4174	1.2836	1.4487
1.65	0.2184	0.3373	0.6475	1.3124	0.4162	1.2922	1.4544
1.66	0.2151	0.3337	0.6447	1.3250	0.4150	1.3010	1.4601
1.67	0.2119	0.3302	0.6419	1.3375	0.4138	1.3100	1.4657
1.68	0.2088	0.3266	0.6392	1.3500	0.4125	1.3190	1.4713
1.69	0.2057	0.3232	0.6364	1.3624	0.4112	1.3283	1.4769
1.70	0.2026	0.3197	0.6337	1.3748	0.4098	1.3376	1.4825
1.71	0.1996	0.3163	0.6310	1.3871	0.4085	1.3471	1.4880
1.72	0.1966	0.3129	0.6283	1.3994	0.4071	1.3567	1.4935
1.73	0.1936	0.3095	0.6256	1.4117	0.4056	1.3665	1.4989
1.74	0.1907	0.3062	0.6229	1.4239	0.4041	1.3764	1.5043
1.75	0.1878	0.3029	0.6202	1.4361	0.4026	1.3865	1.5097
1.76	0.1850	0.2996	0.6175	1.4483	0.4011	1.3967	1.5150
1.77	0.1822	0.2964	0.6148	1.4604	0.3996	1.4070	1.5203
1.78	0.1794	0.2931	0.6121	1.4725	0.3980	1.4175	1.5256

Table A4.2 *Continued*

M	P/P_o	ρ/ρ_o	T/T_o	β	q/P_o	A/A^*	V/a^*
1.79	0.1767	0.2900	0.6095	1.4846	0.3964	1.4282	1.5308
1.80	0.1740	0.2868	0.6068	1.4967	0.3947	1.4390	1.5360
1.81	0.1714	0.2837	0.6041	1.5087	0.3931	1.4499	1.5411
1.82	0.1688	0.2806	0.6015	1.5207	0.3914	1.4610	1.5463
1.83	0.1662	0.2776	0.5989	1.5326	0.3897	1.4723	1.5514
1.84	0.1637	0.2745	0.5963	1.5445	0.3879	1.4836	1.5564
1.85	0.1612	0.2715	0.5936	1.5564	0.3862	1.4952	1.5614
1.86	0.1587	0.2686	0.5910	1.5683	0.3844	1.5069	1.5664
1.87	0.1563	0.2656	0.5884	1.5802	0.3826	1.5187	1.5714
1.88	0.1539	0.2627	0.5859	1.5920	0.3808	1.5308	1.5763
1.89	0.1516	0.2598	0.5833	1.6038	0.3790	1.5429	1.5812
1.90	0.1492	0.2570	0.5807	1.6155	0.3771	1.5553	1.5861
1.91	0.1470	0.2542	0.5782	1.6273	0.3753	1.5677	1.5909
1.92	0.1447	0.2514	0.5756	1.6390	0.3734	1.5804	1.5957
1.93	0.1425	0.2486	0.5731	1.6507	0.3715	1.5932	1.6005
1.94	0.1403	0.2459	0.5705	1.6624	0.3696	1.6062	1.6052
1.95	0.1381	0.2432	0.5680	1.6741	0.3677	1.6193	1.6099
1.96	0.1360	0.2405	0.5655	1.6857	0.3657	1.6326	1.6146
1.97	0.1339	0.2378	0.5630	1.6973	0.3638	1.6461	1.6192
1.98	0.1318	0.2352	0.5605	1.7089	0.3618	1.6597	1.6239
1.99	0.1298	0.2326	0.5580	1.7205	0.3598	1.6735	1.6284
2.00	0.1278	0.2300	0.5556	1.7321	0.3579	1.6875	1.6330
2.01	0.1258	0.2275	0.5531	1.7436	0.3559	1.7016	1.6375
2.02	0.1239	0.2250	0.5506	1.7551	0.3539	1.7160	1.6420
2.03	0.1220	0.2225	0.5482	1.7666	0.3518	1.7305	1.6465
2.04	0.1201	0.2200	0.5458	1.7781	0.3498	1.7451	1.6509
2.05	0.1182	0.2176	0.5433	1.7896	0.3478	1.7600	1.6553
2.06	0.1164	0.2152	0.5409	1.8010	0.3458	1.7750	1.6597
2.07	0.1146	0.2128	0.5385	1.8124	0.3437	1.7902	1.6640
2.08	0.1128	0.2104	0.5361	1.8238	0.3417	1.8056	1.6683
2.09	0.1111	0.2081	0.5337	1.8352	0.3396	1.8212	1.6726
2.10	0.1094	0.2058	0.5313	1.8466	0.3376	1.8369	1.6769
2.11	0.1077	0.2035	0.5290	1.8580	0.3355	1.8529	1.6811
2.12	0.1060	0.2013	0.5266	1.8693	0.3334	1.8690	1.6853
2.13	0.1043	0.1990	0.5243	1.8807	0.3314	1.8853	1.6895
2.14	0.1027	0.1968	0.5219	1.8920	0.3293	1.9018	1.6936
2.15	0.1011	0.1946	0.5196	1.9033	0.3272	1.9185	1.6977
2.16	9.956e–2	0.1925	0.5173	1.9146	0.3252	1.9354	1.7018
2.17	9.802e–2	0.1903	0.5150	1.9259	0.3231	1.9525	1.7059
2.18	9.649e–2	0.1882	0.5127	1.9371	0.3210	1.9698	1.7099
2.19	9.500e–2	0.1861	0.5104	1.9484	0.3189	1.9873	1.7139
2.20	9.352e–2	0.1841	0.5081	1.9596	0.3169	2.0050	1.7179
2.21	9.207e–2	0.1820	0.5059	1.9708	0.3148	2.0229	1.7219
2.22	9.064e–2	0.1800	0.5036	1.9820	0.3127	2.0409	1.7258
2.23	8.923e–2	0.1780	0.5014	1.9932	0.3106	2.0592	1.7297
2.24	8.785e–2	0.1760	0.4991	2.0044	0.3085	2.0777	1.7336
2.25	8.648e–2	0.1740	0.4969	2.0156	0.3065	2.0964	1.7374
2.26	8.514e–2	0.1721	0.4947	2.0267	0.3044	2.1153	1.7412
2.27	8.382e–2	0.1702	0.4925	2.0379	0.3023	2.1345	1.7450
2.28	8.251e–2	0.1683	0.4903	2.0490	0.3003	2.1538	1.7488
2.29	8.123e–2	0.1664	0.4881	2.0601	0.2982	2.1734	1.7526
2.30	7.997e–2	0.1646	0.4859	2.0712	0.2961	2.1931	1.7563
2.31	7.873e–2	0.1628	0.4837	2.0823	0.2941	2.2131	1.7600
2.32	7.751e–2	0.1609	0.4816	2.0934	0.2920	2.2333	1.7637
2.33	7.631e–2	0.1592	0.4794	2.1045	0.2900	2.2538	1.7673
2.34	7.512e–2	0.1574	0.4773	2.1156	0.2879	2.2744	1.7709

Table A4.2 ***Continued***

M	P/P_o	ρ/ρ_o	T/T_o	β	q/P_o	A/A^*	V/a^*
2.35	7.396e–2	0.1556	0.4752	2.1266	0.2859	2.2953	1.7745
2.36	7.281e–2	0.1539	0.4731	2.1377	0.2839	2.3164	1.7781
2.37	7.168e–2	0.1522	0.4709	2.1487	0.2818	2.3377	1.7817
2.38	7.057e–2	0.1505	0.4688	2.1597	0.2798	2.3593	1.7852
2.39	6.948e–2	0.1488	0.4668	2.1707	0.2778	2.3811	1.7887
2.40	6.840e–2	0.1472	0.4647	2.1817	0.2758	2.4031	1.7922
2.41	6.734e–2	0.1456	0.4626	2.1927	0.2738	2.4254	1.7956
2.42	6.630e–2	0.1439	0.4606	2.2037	0.2718	2.4479	1.7991
2.43	6.527e–2	0.1424	0.4585	2.2147	0.2698	2.4706	1.8025
2.44	6.426e–2	0.1408	0.4565	2.2257	0.2678	2.4936	1.8059
2.45	6.327e–2	0.1392	0.4544	2.2366	0.2658	2.5168	1.8092
2.46	6.229e–2	0.1377	0.4524	2.2476	0.2639	2.5403	1.8126
2.47	6.133e–2	0.1362	0.4504	2.2585	0.2619	2.5640	1.8159
2.48	6.038e–2	0.1346	0.4484	2.2694	0.2599	2.5880	1.8192
2.49	5.945e–2	0.1332	0.4464	2.2804	0.2580	2.6122	1.8225
2.50	5.853e–2	0.1317	0.4444	2.2913	0.2561	2.6367	1.8257
2.51	5.762e–2	0.1302	0.4425	2.3022	0.2541	2.6615	1.8290
2.52	5.674e–2	0.1288	0.4405	2.3131	0.2522	2.6865	1.8322
2.53	5.586e–2	0.1274	0.4386	2.3240	0.2503	2.7117	1.8354
2.54	5.500e–2	0.1260	0.4366	2.3349	0.2484	2.7372	1.8386
2.55	5.415e–2	0.1246	0.4347	2.3457	0.2465	2.7630	1.8417
2.56	5.332e–2	0.1232	0.4328	2.3566	0.2446	2.7891	1.8448
2.57	5.250e–2	0.1218	0.4309	2.3675	0.2427	2.8154	1.8479
2.58	5.169e–2	0.1205	0.4289	2.3783	0.2409	2.8420	1.8510
2.59	5.090e–2	0.1192	0.4271	2.3892	0.2390	2.8688	1.8541
2.60	5.012e–2	0.1179	0.4252	2.4000	0.2371	2.8960	1.8571
2.61	4.935e–2	0.1166	0.4233	2.4108	0.2353	2.9234	1.8602
2.62	4.859e–2	0.1153	0.4214	2.4217	0.2335	2.9511	1.8632
2.63	4.784e–2	0.1140	0.4196	2.4325	0.2317	2.9791	1.8662
2.64	4.711e–2	0.1128	0.4177	2.4433	0.2298	3.0073	1.8691
2.65	4.639e–2	0.1115	0.4159	2.4541	0.2280	3.0359	1.8721
2.66	4.568e–2	0.1103	0.4141	2.4649	0.2262	3.0647	1.8750
2.67	4.498e–2	0.1091	0.4122	2.4757	0.2245	3.0938	1.8779
2.68	4.429e–2	0.1079	0.4104	2.4864	0.2227	3.1233	1.8808
2.69	4.362e–2	0.1067	0.4086	2.4972	0.2209	3.1530	1.8837
2.70	4.295e–2	0.1056	0.4068	2.5080	0.2192	3.1830	1.8865
2.71	4.229e–2	0.1044	0.4051	2.5187	0.2174	3.2133	1.8894
2.72	4.165e–2	0.1033	0.4033	2.5295	0.2157	3.2440	1.8922
2.73	4.102e–2	0.1022	0.4015	2.5403	0.2140	3.2749	1.8950
2.74	4.039e–2	0.1010	0.3998	2.5510	0.2123	3.3061	1.8978
2.75	3.978e–2	9.994e–2	0.3980	2.5617	0.2106	3.3377	1.9005
2.76	3.917e–2	9.885e–2	0.3963	2.5725	0.2089	3.3695	1.9033
2.77	3.858e–2	9.778e–2	0.3945	2.5832	0.2072	3.4017	1.9060
2.78	3.799e–2	9.671e–2	0.3928	2.5939	0.2055	3.4342	1.9087
2.79	3.742e–2	9.566e–2	0.3911	2.6046	0.2039	3.4670	1.9114
2.80	3.685e–2	9.463e–2	0.3894	2.6153	0.2022	3.5001	1.9140
2.81	3.629e–2	9.360e–2	0.3877	2.6260	0.2006	3.5336	1.9167
2.82	3.574e–2	9.259e–2	0.3860	2.6367	0.1990	3.5674	1.9193
2.83	3.520e–2	9.158e–2	0.3844	2.6474	0.1973	3.6015	1.9219
2.84	3.467e–2	9.059e–2	0.3827	2.6581	0.1957	3.6359	1.9246
2.85	3.415e–2	8.962e–2	0.3810	2.6688	0.1941	3.6707	1.9271
2.86	3.363e–2	8.865e–2	0.3794	2.6795	0.1926	3.7058	1.9297
2.87	3.312e–2	8.769e–2	0.3777	2.6901	0.1910	3.7413	1.9323
2.88	3.263e–2	8.675e–2	0.3761	2.7008	0.1894	3.7771	1.9348
2.89	3.213e–2	8.581e–2	0.3745	2.7115	0.1879	3.8133	1.9373
2.90	3.165e–2	8.489e–2	0.3729	2.7221	0.1863	3.8498	1.9398

Table A4.2 ***Continued***

M	P/P_o	ρ/ρ_o	T/T_o	β	q/P_o	A/A^*	V/a^*
2.91	3.118e–2	8.398e–2	0.3712	2.7328	0.1848	3.8866	1.9423
2.92	3.071e–2	8.307e–2	0.3696	2.7434	0.1833	3.9238	1.9448
2.93	3.025e–2	8.218e–2	0.3681	2.7541	0.1818	3.9614	1.9472
2.94	2.980e–2	8.130e–2	0.3665	2.7647	0.1803	3.9993	1.9497
2.95	2.935e–2	8.043e–2	0.3649	2.7753	0.1788	4.0376	1.9521
2.96	2.891e–2	7.957e–2	0.3633	2.7860	0.1773	4.0763	1.9545
2.97	2.848e–2	7.872e–2	0.3618	2.7966	0.1758	4.1153	1.9569
2.98	2.805e–2	7.788e–2	0.3602	2.8072	0.1744	4.1547	1.9593
2.99	2.764e–2	7.705e–2	0.3587	2.8178	0.1729	4.1944	1.9616
3.00	2.722e–2	7.623e–2	0.3571	2.8284	0.1715	4.2346	1.9640
3.02	2.642e–2	7.461e–2	0.3541	2.8496	0.1687	4.3160	1.9686
3.04	2.564e–2	7.303e–2	0.3511	2.8708	0.1659	4.3989	1.9732
3.06	2.489e–2	7.149e–2	0.3481	2.8920	0.1631	4.4835	1.9777
3.08	2.416e–2	6.999e–2	0.3452	2.9131	0.1604	4.5696	1.9822
3.10	2.345e–2	6.852e–2	0.3422	2.9343	0.1577	4.6573	1.9866
3.12	2.276e–2	6.708e–2	0.3393	2.9554	0.1551	4.7467	1.9910
3.14	2.210e–2	6.568e–2	0.3365	2.9765	0.1525	4.8377	1.9953
3.16	2.146e–2	6.430e–2	0.3337	2.9976	0.1500	4.9304	1.9995
3.18	2.083e–2	6.296e–2	0.3309	3.0187	0.1475	5.0248	2.0037
3.20	2.023e–2	6.165e–2	0.3281	3.0397	0.1450	5.1210	2.0079
3.22	1.964e–2	6.037e–2	0.3253	3.0608	0.1426	5.2189	2.0119
3.24	1.908e–2	5.912e–2	0.3226	3.0818	0.1402	5.3186	2.0160
3.26	1.853e–2	5.790e–2	0.3199	3.1028	0.1378	5.4201	2.0200
3.28	1.799e–2	5.671e–2	0.3173	3.1238	0.1355	5.5234	2.0239
3.30	1.748e–2	5.554e–2	0.3147	3.1448	0.1332	5.6286	2.0278
3.32	1.698e–2	5.440e–2	0.3121	3.1658	0.1310	5.7358	2.0317
3.34	1.649e–2	5.329e–2	0.3095	3.1868	0.1288	5.8448	2.0355
3.36	1.602e–2	5.220e–2	0.3069	3.2077	0.1266	5.9558	2.0392
3.38	1.557e–2	5.113e–2	0.3044	3.2287	0.1245	6.0687	2.0429
3.40	1.512e–2	5.009e–2	0.3019	3.2496	0.1224	6.1837	2.0466
3.42	1.470e–2	4.908e–2	0.2995	3.2705	0.1203	6.3007	2.0502
3.44	1.428e–2	4.808e–2	0.2970	3.2914	0.1183	6.4198	2.0537
3.46	1.388e–2	4.711e–2	0.2946	3.3123	0.1163	6.5409	2.0573
3.48	1.349e–2	4.616e–2	0.2922	3.3332	0.1144	6.6642	2.0607
3.50	1.311e–2	4.523e–2	0.2899	3.3541	0.1124	6.7896	2.0642
3.52	1.274e–2	4.433e–2	0.2875	3.3750	0.1105	6.9172	2.0676
3.54	1.239e–2	4.344e–2	0.2852	3.3958	0.1087	7.0471	2.0709
3.56	1.204e–2	4.257e–2	0.2829	3.4167	0.1068	7.1791	2.0743
3.58	1.171e–2	4.172e–2	0.2806	3.4375	0.1050	7.3135	2.0775
3.60	1.138e–2	4.089e–2	0.2784	3.4583	0.1033	7.4501	2.0808
3.62	1.107e–2	4.008e–2	0.2762	3.4791	0.1015	7.5891	2.0840
3.64	1.076e–2	3.929e–2	0.2740	3.4999	9.984e–2	7.7305	2.0871
3.66	1.047e–2	3.852e–2	0.2718	3.5207	9.816e–2	7.8742	2.0903
3.68	1.018e–2	3.776e–2	0.2697	3.5415	9.652e–2	8.0204	2.0933
3.70	9.903e–3	3.702e–2	0.2675	3.5623	9.490e–2	8.1691	2.0964
3.72	9.633e–3	3.629e–2	0.2654	3.5831	9.331e–2	8.3202	2.0994
3.74	9.370e–3	3.558e–2	0.2633	3.6038	9.175e–2	8.4739	2.1024
3.76	9.116e–3	3.489e–2	0.2613	3.6246	9.021e–2	8.6302	2.1053
3.78	8.869e–3	3.421e–2	0.2592	3.6453	8.870e–2	8.7891	2.1082
3.80	8.629e–3	3.355e–2	0.2572	3.6661	8.722e–2	8.9506	2.1111
3.82	8.396e–3	3.290e–2	0.2552	3.6868	8.577e–2	9.1148	2.1140
3.84	8.171e–3	3.227e–2	0.2532	3.7075	8.434e–2	9.2817	2.1168
3.86	7.951e–3	3.165e–2	0.2513	3.7282	8.293e–2	9.4513	2.1195
3.88	7.739e–3	3.104e–2	0.2493	3.7489	8.155e–2	9.6237	2.1223
3.90	7.532e–3	3.044e–2	0.2474	3.7696	8.019e–2	9.7990	2.1250
3.92	7.332e–3	2.986e–2	0.2455	3.7903	7.886e–2	9.9771	2.1277

Table A4.2 ***Continued***

M	P/P_o	ρ/ρ_o	T/T_o	β	q/P_o	A/A^*	V/a^*
3.94	7.137e–3	2.929e–2	0.2436	3.8110	7.755e–2	10.158	2.1303
3.96	6.948e–3	2.874e–2	0.2418	3.8317	7.627e–2	10.342	2.1329
3.98	6.764e–3	2.819e–2	0.2399	3.8523	7.500e–2	10.528	2.1355
4.00	6.586e–3	2.766e–2	0.2381	3.8730	7.376e–2	10.718	2.1381
4.04	6.245e–3	2.663e–2	0.2345	3.9143	7.135e–2	11.107	2.1431
4.08	5.923e–3	2.564e–2	0.2310	3.9556	6.902e–2	11.509	2.1480
4.12	5.619e–3	2.470e–2	0.2275	3.9968	6.677e–2	11.923	2.1529
4.16	5.333e–3	2.379e–2	0.2242	4.0380	6.460e–2	12.350	2.1576
4.20	5.062e–3	2.292e–2	0.2208	4.0792	6.251e–2	12.791	2.1622
4.24	4.806e–3	2.209e–2	0.2176	4.1204	6.049e–2	13.246	2.1667
4.28	4.565e–3	2.129e–2	0.2144	4.1615	5.854e–2	13.715	2.1711
4.32	4.337e–3	2.052e–2	0.2113	4.2027	5.666e–2	14.198	2.1754
4.36	4.121e–3	1.979e–2	0.2083	4.2438	5.484e–2	14.696	2.1796
4.40	3.918e–3	1.909e–2	0.2053	4.2849	5.309e–2	15.209	2.1837
4.44	3.725e–3	1.841e–2	0.2023	4.3259	5.140e–2	15.738	2.1877
4.48	3.543e–3	1.776e–2	0.1994	4.3670	4.977e–2	16.283	2.1917
4.52	3.370e–3	1.714e–2	0.1966	4.4080	4.820e–2	16.844	2.1955
4.56	3.207e–3	1.654e–2	0.1938	4.4490	4.668e–2	17.422	2.1993
4.60	3.053e–3	1.597e–2	0.1911	4.4900	4.521e–2	18.017	2.2030
4.64	2.906e–3	1.542e–2	0.1885	4.5310	4.380e–2	18.630	2.2066
4.68	2.768e–3	1.489e–2	0.1859	4.5719	4.243e–2	19.260	2.2102
4.72	2.637e–3	1.438e–2	0.1833	4.6129	4.112e–2	19.909	2.2136
4.76	2.512e–3	1.390e–2	0.1808	4.6538	3.984e–2	20.577	2.2170
4.80	2.394e–3	1.343e–2	0.1783	4.6947	3.861e–2	21.263	2.2204
4.84	2.283e–3	1.298e–2	0.1759	4.7356	3.743e–2	21.970	2.2236
4.88	2.177e–3	1.254e–2	0.1735	4.7764	3.628e–2	22.696	2.2268
4.92	2.076e–3	1.213e–2	0.1712	4.8173	3.518e–2	23.443	2.2300
4.96	1.981e–3	1.173e–2	0.1689	4.8581	3.411e–2	24.210	2.2331
5.00	1.890e–3	1.134e–2	0.1667	4.8990	3.308e–2	25.000	2.2361
5.10	1.683e–3	1.044e–2	0.1612	5.0010	3.065e–2	27.069	2.2433
5.20	1.501e–3	9.620e–3	0.1561	5.1029	2.842e–2	29.283	2.2503
5.30	1.341e–3	8.875e–3	0.1511	5.2048	2.637e–2	31.649	2.2569
5.40	1.200e–3	8.197e–3	0.1464	5.3066	2.449e–2	34.174	2.2631
5.50	1.075e–3	7.578e–3	0.1418	5.4083	2.276e–2	36.869	2.2691
5.60	9.643e–4	7.012e–3	0.1375	5.5100	2.117e–2	39.740	2.2748
5.70	8.663e–4	6.496e–3	0.1334	5.6116	1.970e–2	42.797	2.2803
5.80	7.794e–4	6.023e–3	0.1294	5.7131	1.835e–2	46.050	2.2855
5.90	7.021e–4	5.590e–3	0.1256	5.8146	1.711e–2	49.507	2.2905
6.00	6.334e–4	5.194e–3	0.1220	5.9161	1.596e–2	53.179	2.2953

Appendix 5: Shock wave data

Table A5.1 Normal shock wave data

Pressure, Mach number and temperature changes through shock waves ($\gamma = 7/5$).

Notation:
M_1 = Mach number of flow upstream of shock wave
M_2 = Mach number of flow behind the shock wave
ν = Prandtl–Meyer angle, (deg), for expanding flow at M_1
μ = Mach angle, (deg), $(\sin(-1)(1/M_1))$
P_2/P_1 = Static pressure ratio across normal shock wave
d_2/d_1 = Density ratio across normal shock wave
T_2/T_1 = Temperature ratio across normal shock wave
P_{o2}/P_{o1} = Stagnation pressure ratio across normal shock wave

M_1	ν	μ	M_2	P_2/P_1	d_2/d_1	T_2/T_1	P_{o2}/P_{o1}
1.00	0.000	90.000	1.0000	1.000	1.0000	1.0000	1.0000
1.01	0.045	81.931	0.9901	1.023	1.0167	1.0066	1.0000
1.02	0.126	78.635	0.9805	1.047	1.0334	1.0132	1.0000
1.03	0.229	76.138	0.9712	1.071	1.0502	1.0198	1.0000
1.04	0.351	74.058	0.9620	1.095	1.0671	1.0263	0.9999
1.05	0.487	72.247	0.9531	1.120	1.0840	1.0328	0.9999
1.06	0.637	70.630	0.9444	1.144	1.1009	1.0393	0.9998
1.07	0.797	69.160	0.9360	1.169	1.1179	1.0458	0.9996
1.08	0.968	67.808	0.9277	1.194	1.1349	1.0522	0.9994
1.09	1.148	66.553	0.9196	1.219	1.1520	1.0586	0.9992
1.10	1.336	65.380	0.9118	1.245	1.1691	1.0649	0.9989
1.11	1.532	64.277	0.9041	1.271	1.1862	1.0713	0.9986
1.12	1.735	63.234	0.8966	1.297	1.2034	1.0776	0.9982
1.13	1.944	62.246	0.8892	1.323	1.2206	1.0840	0.9978
1.14	2.160	61.306	0.8820	1.350	1.2378	1.0903	0.9973
1.15	2.381	60.408	0.8750	1.376	1.2550	1.0966	0.9967
1.16	2.607	59.550	0.8682	1.403	1.2723	1.1029	0.9961
1.17	2.839	58.727	0.8615	1.430	1.2896	1.1092	0.9953
1.18	3.074	57.936	0.8549	1.458	1.3069	1.1154	0.9946
1.19	3.314	57.176	0.8485	1.485	1.3243	1.1217	0.9937
1.20	3.558	56.443	0.8422	1.513	1.3416	1.1280	0.9928
1.21	3.806	55.735	0.8360	1.541	1.3590	1.1343	0.9918
1.22	4.057	55.052	0.8300	1.570	1.3764	1.1405	0.9907
1.23	4.312	54.391	0.8241	1.598	1.3938	1.1468	0.9896
1.24	4.569	53.751	0.8183	1.627	1.4112	1.1531	0.9884
1.25	4.830	53.130	0.8126	1.656	1.4286	1.1594	0.9871

Table A5.1 ***Continued***

M_1	ν	μ	M_2	P_2/P_1	d_2/d_1	T_2/T_1	P_{o2}/P_{o1}
1.26	5.093	52.528	0.8071	1.686	1.4460	1.1657	0.9857
1.27	5.359	51.943	0.8016	1.715	1.4634	1.1720	0.9842
1.28	5.627	51.375	0.7963	1.745	1.4808	1.1783	0.9827
1.29	5.898	50.823	0.7911	1.775	1.4983	1.1846	0.9811
1.30	6.170	50.285	0.7860	1.805	1.5157	1.1909	0.9794
1.31	6.445	49.761	0.7809	1.835	1.5331	1.1972	0.9776
1.32	6.721	49.251	0.7760	1.866	1.5505	1.2035	0.9758
1.33	7.000	48.753	0.7712	1.897	1.5680	1.2099	0.9738
1.34	7.279	48.268	0.7664	1.928	1.5854	1.2162	0.9718
1.35	7.561	47.795	0.7618	1.960	1.6028	1.2226	0.9697
1.36	7.844	47.332	0.7572	1.991	1.6202	1.2290	0.9676
1.37	8.128	46.880	0.7527	2.023	1.6376	1.2354	0.9653
1.38	8.413	46.439	0.7483	2.055	1.6549	1.2418	0.9630
1.39	8.699	46.007	0.7440	2.087	1.6723	1.2482	0.9607
1.40	8.987	45.585	0.7397	2.120	1.6897	1.2547	0.9582
1.41	9.276	45.171	0.7355	2.153	1.7070	1.2612	0.9557
1.42	9.565	44.767	0.7314	2.186	1.7243	1.2676	0.9531
1.43	9.855	44.371	0.7274	2.219	1.7416	1.2741	0.9504
1.44	10.146	43.983	0.7235	2.253	1.7589	1.2807	0.9473
1.45	10.438	43.603	0.7196	2.286	1.7761	1.2872	0.9448
1.46	10.731	43.230	0.7157	2.320	1.7934	1.2938	0.9420
1.47	11.023	42.865	0.7120	2.354	1.8106	1.3003	0.9390
1.48	11.317	42.507	0.7083	2.389	1.8278	1.3069	0.9360
1.49	11.611	42.155	0.7047	2.423	1.8449	1.3136	0.9329
1.50	11.905	41.810	0.7011	2.458	1.8621	1.3202	0.9298
1.51	12.200	41.472	0.6976	2.493	1.8792	1.3269	0.9266
1.52	12.495	41.140	0.6941	2.529	1.8963	1.3336	0.9233
1.53	12.790	40.813	0.6907	2.564	1.9133	1.3403	0.9200
1.54	13.086	40.493	0.6874	2.600	1.9303	1.3470	0.9166
1.55	13.381	40.178	0.6841	2.636	1.9473	1.3538	0.9132
1.56	13.677	39.868	0.6809	2.673	1.9643	1.3606	0.9097
1.57	13.973	39.564	0.6777	2.709	1.9812	1.3674	0.9062
1.58	14.269	39.265	0.6746	2.746	1.9981	1.3742	0.9026
1.59	14.565	38.971	0.6715	2.783	2.0149	1.3811	0.8989
1.60	14.860	38.682	0.6684	2.820	2.0317	1.3880	0.8952
1.61	15.156	38.398	0.6655	2.857	2.0485	1.3949	0.8915
1.62	15.452	38.118	0.6625	2.895	2.0653	1.4018	0.8877
1.63	15.747	37.843	0.6596	2.933	2.0820	1.4088	0.8838
1.64	16.043	37.572	0.6568	2.971	2.0986	1.4158	0.8799
1.65	16.338	37.305	0.6540	3.010	2.1152	1.4228	0.8760
1.66	16.633	37.043	0.6512	3.048	2.1318	1.4299	0.8720
1.67	16.928	36.784	0.6485	3.087	2.1484	1.4369	0.8680
1.68	17.222	36.530	0.6458	3.126	2.1649	1.4440	0.8639
1.69	17.516	36.279	0.6431	3.165	2.1813	1.4512	0.8599
1.70	17.810	36.032	0.6405	3.205	2.1977	1.4583	0.8557
1.71	18.103	35.789	0.6380	3.245	2.2141	1.4655	0.8516
1.72	18.396	35.549	0.6355	3.285	2.2304	1.4727	0.8474
1.73	18.689	35.312	0.6330	3.325	2.2467	1.4800	0.8431
1.74	18.981	35.080	0.6305	3.366	2.2629	1.4873	0.8389
1.75	19.273	34.850	0.6281	3.406	2.2791	1.4946	0.8346
1.76	19.565	34.624	0.6257	3.447	2.2952	1.5019	0.8302
1.77	19.855	34.400	0.6234	3.488	2.3113	1.5093	0.8259
1.78	20.146	34.180	0.6210	3.530	2.3273	1.5167	0.8215
1.79	20.436	33.963	0.6188	3.571	2.3433	1.5241	0.8171
1.80	20.725	33.749	0.6165	3.613	2.3592	1.5316	0.8127
1.81	21.014	33.538	0.6143	3.655	2.3751	1.5391	0.8082

Table A5.1 ***Continued***

M_1	ν	μ	M_2	P_2/P_1	d_2/d_1	T_2/T_1	P_{o2}/P_{o1}
1.82	21.302	33.329	0.6121	3.698	2.3909	1.5466	0.8038
1.83	21.590	33.124	0.6099	3.740	2.4067	1.5541	0.7993
1.84	21.877	32.921	0.6078	3.783	2.4224	1.5617	0.7948
1.85	22.163	32.720	0.6057	3.826	2.4381	1.5693	0.7902
1.86	22.449	32.523	0.6036	3.870	2.4537	1.5770	0.7857
1.87	22.734	32.328	0.6016	3.913	2.4693	1.5847	0.7811
1.88	23.019	32.135	0.5996	3.957	2.4848	1.5924	0.7765
1.89	23.303	31.945	0.5976	4.001	2.5003	1.6001	0.7720
1.90	23.586	31.757	0.5956	4.045	2.5157	1.6079	0.7674
1.91	23.869	31.571	0.5937	4.089	2.5310	1.6157	0.7627
1.92	24.151	31.388	0.5918	4.134	2.5463	1.6236	0.7581
1.93	24.432	31.207	0.5899	4.179	2.5616	1.6314	0.7535
1.94	24.712	31.028	0.5880	4.224	2.5767	1.6394	0.7488
1.95	24.992	30.852	0.5862	4.270	2.5919	1.6473	0.7442
1.96	25.271	30.677	0.5844	4.315	2.6069	1.6553	0.7395
1.97	25.549	30.505	0.5826	4.361	2.6220	1.6633	0.7349
1.98	25.827	30.335	0.5808	4.407	2.6369	1.6713	0.7302
1.99	26.104	30.166	0.5791	4.453	2.6518	1.6794	0.7255
2.00	26.380	30.000	0.5774	4.500	2.6667	1.6875	0.7209
2.01	26.655	29.836	0.5757	4.547	2.6815	1.6956	0.7162
2.02	26.930	29.673	0.5740	4.594	2.6962	1.7038	0.7115
2.03	27.203	29.512	0.5723	4.641	2.7109	1.7120	0.7069
2.04	27.476	29.353	0.5707	4.689	2.7255	1.7203	0.7022
2.05	27.748	29.196	0.5691	4.736	2.7400	1.7285	0.6975
2.06	28.020	29.041	0.5675	4.784	2.7545	1.7369	0.6928
2.07	28.290	28.888	0.5659	4.832	2.7689	1.7452	0.6882
2.08	28.560	28.736	0.5643	4.881	2.7833	1.7536	0.6835
2.09	28.829	28.585	0.5628	4.929	2.7976	1.7620	0.6789
2.10	29.097	28.437	0.5613	4.978	2.8119	1.7705	0.6742
2.11	29.364	28.290	0.5598	5.027	2.8261	1.7789	0.6696
2.12	29.631	28.145	0.5583	5.077	2.8402	1.7875	0.6649
2.13	29.896	28.001	0.5568	5.126	2.8543	1.7960	0.6603
2.14	30.161	27.859	0.5554	5.176	2.8683	1.8046	0.6557
2.15	30.425	27.718	0.5540	5.226	2.8823	1.8132	0.6511
2.16	30.688	27.578	0.5525	5.277	2.8962	1.8219	0.6464
2.17	30.951	27.441	0.5511	5.327	2.9101	1.8306	0.6419
2.18	31.212	27.304	0.5498	5.378	2.9238	1.8393	0.6373
2.19	31.473	27.169	0.5484	5.429	2.9376	1.8481	0.6327
2.20	21.732	27.036	0.5471	5.480	2.9512	1.8569	0.6281
2.21	31.991	26.903	0.5457	5.531	2.9648	1.8657	0.6236
2.22	32.249	26.773	0.5444	5.583	2.9784	1.8746	0.6191
2.23	32.507	26.643	0.5431	5.635	2.9918	1.8835	0.6145
2.24	32.763	26.515	0.5418	5.687	3.0053	1.8924	0.6100
2.25	33.018	26.388	0.5406	5.740	3.0186	1.9014	0.6055
2.26	33.273	26.262	0.5393	5.792	3.0319	1.9104	0.6011
2.27	33.527	26.138	0.5381	5.845	3.0452	1.9194	0.5966
2.28	33.780	26.014	0.5368	5.898	3.0584	1.9285	0.5921
2.29	34.032	25.892	0.5356	5.951	3.0715	1.9376	0.5877
2.30	34.283	25.771	0.5344	6.005	3.0845	1.9468	0.5833
2.31	34.533	25.652	0.5332	6.059	3.0976	1.9560	0.5789
2.32	34.782	25.533	0.5321	6.113	3.1105	1.9652	0.5745
2.33	35.031	25.416	0.5309	6.167	3.1234	1.9745	0.5702
2.34	35.279	25.300	0.5297	6.222	3.1362	1.9838	0.5658
2.35	35.526	25.184	0.5286	6.276	3.1490	1.9931	0.5615
2.36	35.771	25.070	0.5275	6.331	3.1617	2.0025	0.5572
2.37	36.017	24.957	0.5264	6.386	3.1743	2.0119	0.5529

Table A5.1 ***Continued***

M_1	ν	μ	M_2	P_2/P_1	d_2/d_1	T_2/T_1	P_{o2}/P_{o1}
2.38	36.261	24.845	0.5253	6.442	3.1869	2.0213	0.5486
2.39	36.504	24.734	0.5242	6.497	3.1994	2.0308	0.5444
2.40	36.747	24.624	0.5231	6.553	3.2119	2.0403	0.5401
2.41	36.988	24.515	0.5221	6.609	3.2243	2.0499	0.5359
2.42	37.229	24.407	0.5210	6.666	3.2367	2.0595	0.5317
2.43	37.469	24.301	0.5200	6.722	3.2489	2.0691	0.5276
2.44	37.708	24.195	0.5189	6.779	3.2612	2.0788	0.5234
2.45	37.946	24.090	0.5179	6.836	3.2733	2.0885	0.5193
2.46	38.183	23.985	0.5169	6.894	3.2855	2.0982	0.5152
2.47	38.420	23.882	0.5159	6.951	3.2975	2.1080	0.5111
2.48	38.655	23.780	0.5149	7.009	3.3095	2.1178	0.5071
2.49	38.890	23.679	0.5140	7.067	3.3215	2.1276	0.5030
2.50	39.124	23.578	0.5130	7.125	3.3333	2.1375	0.4990
2.51	39.357	23.479	0.5120	7.183	3.3452	2.1474	0.4950
2.52	39.589	23.380	0.5111	7.242	3.3569	2.1574	0.4911
2.53	39.820	23.282	0.5102	7.301	3.3686	2.1674	0.4871
2.54	40.050	23.185	0.5092	7.360	3.3803	2.1774	0.4832
2.55	40.280	23.089	0.5083	7.420	3.3919	2.1875	0.4793
2.56	40.508	22.993	0.5074	7.479	3.4034	2.1976	0.4754
2.57	40.736	22.899	0.5065	7.539	3.4149	2.2077	0.4715
2.58	40.963	22.805	0.5056	7.599	3.4263	2.2179	0.4677
2.59	41.189	22.712	0.5047	7.659	3.4377	2.2281	0.4639
2.60	41.415	22.620	0.5039	7.720	3.4490	2.2383	0.4601
2.61	41.639	22.528	0.5030	7.781	3.4602	2.2486	0.4564
2.62	41.863	22.438	0.5022	7.842	3.4714	2.2590	0.4526
2.63	42.086	22.348	0.5013	7.903	3.4826	2.2693	0.4489
2.64	42.307	22.259	0.5005	7.965	3.4937	2.2797	0.4452
2.65	42.529	22.170	0.4996	8.026	3.5047	2.2902	0.4416
2.66	42.749	22.082	0.4988	8.088	3.5157	2.3006	0.4379
2.67	42.968	21.995	0.4980	8.150	3.5266	2.3111	0.4343
2.68	43.187	21.909	0.4972	8.213	3.5374	2.3217	0.4307
2.69	43.405	21.823	0.4964	8.275	3.5482	2.3323	0.4271
2.70	43.621	21.738	0.4956	8.338	3.5590	2.3429	0.4236
2.71	43.838	21.654	0.4949	8.401	3.5697	2.3536	0.4201
2.72	44.053	21.571	0.4941	8.465	3.5803	2.3642	0.4166
2.73	44.267	21.488	0.4933	8.528	3.5909	2.3750	0.4131
2.74	44.481	21.405	0.4926	8.592	3.6015	2.3858	0.4097
2.75	44.694	21.324	0.4918	8.656	3.6119	2.3966	0.4062
2.76	44.906	21.243	0.4911	8.721	3.6224	2.4074	0.4028
2.77	45.117	21.162	0.4903	8.785	3.6327	2.4183	0.3994
2.78	45.327	21.083	0.4896	8.850	3.6431	2.4292	0.3961
2.79	45.537	21.003	0.4889	8.915	3.6533	2.4402	0.3928
2.80	45.746	20.925	0.4882	8.980	3.6636	2.4512	0.3895
2.81	45.954	20.847	0.4875	9.045	3.6737	2.4622	0.3862
2.82	46.161	20.770	0.4868	9.111	3.6838	2.4733	0.3829
2.83	46.368	20.693	0.4861	9.177	3.6939	2.4844	0.3797
2.84	46.573	20.617	0.4854	9.243	3.7039	2.4955	0.3765
2.85	46.778	20.541	0.4847	9.310	3.7139	2.5067	0.3733
2.86	46.982	20.466	0.4840	9.376	3.7238	2.5179	0.3701
2.87	47.185	20.391	0.4833	9.443	3.7336	2.5292	0.3670
2.88	47.388	20.318	0.4827	9.510	3.7434	2.5405	0.3639
2.89	47.589	20.244	0.4820	9.577	3.7532	2.5518	0.3608
2.90	47.790	20.171	0.4814	9.645	3.7629	2.5632	0.3577
2.91	47.990	20.099	0.4807	9.713	3.7725	2.5746	0.3547
2.92	48.190	20.027	0.4801	9.781	3.7821	2.5861	0.3517
2.93	48.388	19.956	0.4795	9.849	3.7917	2.5976	0.3487

Table A5.1 ***Continued***

M_1	ν	μ	M_2	P_2/P_1	d_2/d_1	T_2/T_1	P_{o2}/P_{o1}
2.94	48.586	19.885	0.4788	9.918	3.8012	2.6091	0.3457
2.95	48.783	19.815	0.4782	9.986	3.8106	2.6206	0.3428
2.96	48.980	19.745	0.4776	10.05	3.8200	2.6322	0.3398
2.97	49.175	19.676	0.4770	10.12	3.8294	2.6439	0.3369
2.98	49.370	19.607	0.4764	10.19	3.8387	2.6555	0.3340
2.99	49.564	19.539	0.4758	10.26	3.8479	2.6673	0.3312
3.00	49.757	19.471	0.4752	10.33	3.8571	2.6790	0.3283
3.02	50.142	19.337	0.4740	10.47	3.8754	2.7026	0.3227
3.04	50.523	19.205	0.4729	10.61	3.8935	2.7264	0.3172
3.06	50.902	19.075	0.4717	10.75	3.9114	2.7503	0.3118
3.08	51.277	18.946	0.4706	10.90	3.9291	2.7744	0.3065
3.10	51.650	18.819	0.4695	11.04	3.9466	2.7986	0.3012
3.12	52.020	18.694	0.4685	11.19	3.9639	2.8230	0.2960
3.14	52.386	18.571	0.4674	11.33	3.9811	2.8475	0.2910
3.16	52.751	18.449	0.4664	11.48	3.9981	2.8722	0.2860
3.18	53.112	18.329	0.4654	11.63	4.0149	2.8970	0.2811
3.20	53.470	18.210	0.4643	11.78	4.0315	2.9220	0.2762
3.22	53.826	18.093	0.4634	11.93	4.0479	2.9471	0.2715
3.24	54.179	17.977	0.4624	12.08	4.0642	2.9724	0.2668
3.26	54.529	17.863	0.4614	12.23	4.0803	2.9979	0.2622
3.28	54.877	17.751	0.4605	12.38	4.0963	3.0234	0.2577
3.30	55.222	17.640	0.4596	12.53	4.1120	3.0492	0.2533
3.32	55.564	17.530	0.4587	12.69	4.1276	3.0751	0.2489
3.34	55.904	17.422	0.4578	12.84	4.1431	3.1011	0.2446
3.36	56.241	17.315	0.4569	13.00	4.1583	3.1273	0.2404
3.38	56.576	17.209	0.4560	13.16	4.1734	3.1537	0.2363
3.40	56.908	17.105	0.4552	13.32	4.1884	3.1802	0.2322
3.42	57.237	17.002	0.4544	13.47	4.2032	3.2069	0.2282
3.44	57.564	16.900	0.4535	13.63	4.2179	3.2337	0.2243
3.46	57.888	16.799	0.4527	13.80	4.2323	3.2607	0.2205
3.48	58.210	16.700	0.4519	13.96	4.2467	3.2878	0.2167
3.50	58.530	16.602	0.4512	14.12	4.2609	3.3151	0.2129
3.52	58.847	16.505	0.4504	14.28	4.2749	3.3425	0.2093
3.54	59.162	16.409	0.4496	14.45	4.2888	3.3701	0.2057
3.56	59.474	16.314	0.4489	14.61	4.3026	3.3978	0.2022
3.58	59.784	16.220	0.4481	14.78	4.3162	3.4257	0.1987
3.60	60.091	16.128	0.4474	14.95	4.3296	3.4537	0.1953
3.62	60.397	16.036	0.4467	15.12	4.3429	3.4819	0.1920
3.64	60.700	15.946	0.4460	15.29	4.3561	3.5103	0.1887
3.66	61.001	15.856	0.4453	15.46	4.3692	3.5388	0.1855
3.68	61.299	15.768	0.4446	15.63	4.3821	3.5674	0.1823
3.70	61.595	15.680	0.4439	15.80	4.3949	3.5962	0.1792
3.72	61.889	15.594	0.4433	15.97	4.4075	3.6252	0.1761
3.74	62.181	15.508	0.4426	16.15	4.4200	3.6543	0.1731
3.76	62.471	15.424	0.4420	16.32	4.4324	3.6836	0.1702
3.78	62.758	15.340	0.4414	16.50	4.4447	3.7130	0.1673
3.80	63.044	15.258	0.4407	16.68	4.4568	3.7426	0.1645
3.82	63.327	15.176	0.4401	16.85	4.4688	3.7723	0.1617
3.84	63.608	15.095	0.4395	17.03	4.4807	3.8022	0.1589
3.86	63.887	15.015	0.4389	17.21	4.4924	3.8323	0.1563
3.88	64.164	14.936	0.4383	17.39	4.5041	3.8625	0.1536
3.90	64.440	14.857	0.4377	17.57	4.5156	3.8928	0.1510
3.92	64.713	14.780	0.4372	17.76	4.5270	3.9233	0.1485
3.94	64.984	14.703	0.4366	17.94	4.5383	3.9540	0.1460
3.96	65.253	14.627	0.4360	18.12	4.5494	3.9848	0.1435
3.98	65.520	14.552	0.4355	18.31	4.5605	4.0158	0.1411

Table A5.1 ***Continued***

M_1	ν	μ	M_2	P_2/P_1	d_2/d_1	T_2/T_1	P_{o2}/P_{o1}
4.00	65.785	14.478	0.4350	18.50	4.5714	4.0469	0.1388
4.05	66.439	14.295	0.4336	18.97	4.5983	4.1254	0.1330
4.10	67.082	14.117	0.4324	19.44	4.6245	4.2048	0.1276
4.15	67.713	13.943	0.4311	19.92	4.6500	4.2852	0.1223
4.20	68.333	13.774	0.4299	20.41	4.6749	4.3666	0.1173
4.25	68.942	13.609	0.4288	20.90	4.6992	4.4489	0.1126
4.30	69.541	13.448	0.4277	21.40	4.7229	4.5322	0.1080
4.35	70.129	13.290	0.4266	21.91	4.7460	4.6165	0.1036
4.40	70.706	13.137	0.4255	22.42	4.7685	4.7017	9.948e–2
4.45	71.274	12.986	0.4245	22.93	4.7904	4.7879	9.550e–2
4.50	71.832	12.840	0.4236	23.45	4.8119	4.8751	9.170e–2
4.55	72.380	12.696	0.4226	23.98	4.8328	4.9632	8.806e–2
4.60	72.919	12.556	0.4217	24.52	4.8532	5.0523	8.459e–2
4.65	73.449	12.419	0.4208	25.06	4.8731	5.1424	8.126e–2
4.70	73.970	12.284	0.4199	25.60	4.8926	5.2334	7.809e–2
4.75	74.482	12.153	0.4191	26.15	4.9116	5.3254	7.505e–2
4.80	74.986	12.025	0.4183	26.71	4.9301	5.4184	7.214e–2
4.85	75.482	11.899	0.4175	27.27	4.9482	5.5124	6.936e–2
4.90	75.969	11.776	0.4167	27.84	4.9659	5.6073	6.670e–2
4.95	76.449	11.655	0.4160	28.42	4.9831	5.7032	6.415e–2
5.00	76.920	11.537	0.4152	29.00	5.0000	5.8000	6.172e–2
5.10	77.841	11.308	0.4138	30.17	5.0326	5.9966	5.715e–2
5.20	78.732	11.087	0.4125	31.38	5.0637	6.1971	5.297e–2
5.30	79.596	10.876	0.4113	32.60	5.0934	6.4014	4.913e–2
5.40	80.433	10.672	0.4101	33.85	5.1218	6.6097	4.560e–2
5.50	81.245	10.476	0.4090	35.12	5.1489	6.8218	4.236e–2
5.60	82.032	10.287	0.4079	36.42	5.1749	7.0378	3.938e–2
5.70	82.796	10.104	0.4069	37.73	5.1998	7.2577	3.664e–2
5.80	83.537	9.928	0.4059	39.08	5.2236	7.4814	3.412e–2
5.90	84.256	9.758	0.4050	40.44	5.2464	7.7091	3.179e–2
6.00	84.955	9.594	0.4042	41.83	5.2683	7.9406	2.965e–2

Table A5.2 Oblique shock waves (isentropic flow, γ=7/5)

Notation:
M_1 = Upstream flow Mach number
M_2 = Downstream flow Mach number
δ = (Delta) flow deflection angle
θ = (Theta) wave angle
P_2/P_1 = Ratio of static pressures across wave

M_1	δ	*Weak solution*		
		θ	M_2	P_2/P_1
1.05	0.0	72.25	1.050	1.000
1.10	0.0	65.38	1.100	1.000
1.10	1.0	69.81	1.039	1.077
1.15	0.0	60.41	1.150	1.000
1.15	1.0	63.16	1.102	1.062
1.15	2.0	67.01	1.043	1.141

Table A5.2 ***Continued***

M_1	δ	Weak solution		
		θ	M_2	P_2/P_1
1.20	0.0	56.44	1.200	1.000
1.20	1.0	58.55	1.158	1.056
1.20	2.0	61.05	1.111	1.120
1.20	3.0	64.34	1.056	1.198
1.25	0.0	53.13	1.25	1.000
1.25	1.0	54.88	1.211	1.053
1.25	2.0	56.85	1.170	1.111
1.25	3.0	59.13	1.124	1.176
1.25	4.0	61.99	1.072	1.254
1.25	5.0	66.59	0.999	1.366
1.30	0.0	50.29	1.300	1.000
1.30	1.0	51.81	1.263	1.051
1.30	2.0	53.48	1.224	1.107
1.30	3.0	55.32	1.184	1.167
1.30	4.0	57.42	1.140	1.233
1.30	5.0	59.96	1.090	1.311
1.30	6.0	63.46	1.027	1.411
1.35	0.0	47.80	1.350	1.000
1.35	1.0	49.17	1.314	1.051
1.35	2.0	50.64	1.277	1.104
1.35	3.0	52.22	1.239	1.162
1.35	4.0	53.97	1.199	1.224
1.35	5.0	55.93	1.157	1.292
1.35	6.0	58.23	1.109	1.370
1.35	7.0	61.18	1.052	1.466
1.35	8.0	66.92	0.954	1.633
1.40	0.0	45.59	1.400	1.000
1.40	1.0	46.84	1.365	1.050
1.40	2.0	48.17	1.330	1.103
1.40	3.0	49.59	1.293	1.159
1.40	4.0	51.12	1.255	1.219
1.40	5.0	52.78	1.216	1.283
1.40	6.0	54.63	1.174	1.354
1.40	7.0	56.76	1.128	1.433
1.40	8.0	59.37	1.074	1.526
1.40	9.0	63.19	1.003	1.655
2.20	0.0	27.04	2.200	1.000
2.20	2.0	28.59	2.124	1.127
2.20	4.0	30.24	2.049	1.265
2.20	6.0	31.98	1.974	1.417
2.20	8.0	33.83	1.899	1.583
2.20	10.0	35.79	1.823	1.764
2.20	12.0	37.87	1.745	1.961
2.20	14.0	40.10	1.666	2.176
2.20	16.0	42.49	1.583	2.410
2.20	18.0	45.09	1.496	2.666
2.20	20.0	47.98	1.404	2.949
2.20	22.0	51.28	1.301	3.270
2.20	24.0	55.36	1.181	3.655
2.20	26.0	62.70	0.980	4.292

Index

Printed and bound by CPI Group (UK) Ltd, Croydon, CR0 4YY

22/10/2024

01777547-0001